THE
PHYSICS
OF WAR

ALSO BY BARRY PARKER

Einstein's Brainchild

Quantum Legacy

Einstein: The Passions of a Scientist

Albert Einstein's Vision

THE
PHYSICS
OF WAR

$$\left[\begin{array}{c} \text{FROM ARROWS} \\ \text{TO ATOMS} \end{array} \right]$$

BARRY PARKER

Prometheus Books

59 John Glenn Drive
Amherst, New York 14228

Published 2014 by Prometheus Books

Interior artwork by Lori Scoffield Beer
Cover image © 2013 Media Bakery and BIGSTOCK
Jacket design by Grace M. Conti-Zilsberger

Inquiries should be addressed to
Prometheus Books
59 John Glenn Drive
Amherst, New York 14228–2119
VOICE: 716–691–0133
FAX: 716–691–0137
WWW.PROMETHEUSBOOKS.COM

18 17 16 15 14   5 4 3 2 1

Library of Congress Cataloging-in-Publication Data Pending

ISBN 978-1-61614-803-4 (hardcover)
ISBN 978-1-61614-804-1 (ebook)

Printed in the United States of America

CONTENTS

PREFACE

I mentioned to a friend that I was writing a book on the physics of war. "What does physics have to do with war?" he asked. "Oh, you mean the atomic bomb," he added. And indeed most people know that physics had something to do with the atomic bomb. But in reality it has made many contributions other than the atomic bomb, and not all of them have led to the creation of offensive weapons that have caused damage and grief. It has also been helpful for defense, and one of the best examples of this was the invention of radar just before World War II. It allowed the British to track incoming German planes and take action to defend themselves. The invention of radar no doubt saved thousands of lives during the Battle of Britain. The discovery of x-rays by Röntgen has also played a large role in war, and there's no doubt that it has saved lives.

And it's not just modern weapons that have been based on the principles of physics. Even though they knew little physics, early civilizations, such as the Egyptians, Assyrians, Greeks and Romans, used physics in devising their weapons. Indeed, all through history physics has played an important role in the development of weapons.

As the basic principles of physics were discovered by such people as Galileo, Newton, Huygens, Einstein, and others, it became a science that was solidly based on a firm foundation. At the same time, however, it became more complex and more difficult for the layperson to understand. But it is important that people other than scientists understand, at least to some degree, what is going on in the world of science, and I'm hoping that the present book will help in this respect. I would like to also mention that although physics has been used extensively in war, it has been found to have many applications for the betterment of humankind.

In as many places as possible I've tried to use a story format to make the book more readable and interesting. I've used a few formulas; I hope they don't scare you. I've added them for anyone that is particularly interested in the details of the physics behind the weapons. You can ignore them without losing much.

Finally, I would like to thank my artist, Lori Beer, for an excellent job on the diagrams. I'm sure they will be helpful to most people.

1
INTRODUCTION

The first well-documented battle in the history of the world took place in 1457 BCE on the Plain of Esdraelon, near the city of Megiddo in modern day Syria. It is usually referred to as the Battle of Megiddo. Megiddo, along with several other cities in the region of Palestine and Syria, had formed a coalition under the Prince of Kadesh, and had decided to break away from Egypt. Egypt's pharaoh, Thutmose III, was determined to stop the rebellion. With an army of ten thousand to fifteen thousand men, including infantry, archers, and cavalry, he marched toward Megiddo, arriving within a few miles of it in April. As the army camped at a place called Yaham, Thutmose conferred with his generals. There were three routes from Yaham to Megiddo; two were relatively easy, but a third, more direct route through the mountains, was quite difficult. Part of the route passed through a very narrow pass where his soldiers would have to travel single file. In addition, the cavalry would have to dismount and lead their horses. Strung out in this way, they would be vulnerable if the Prince of Kadesh decided to attack. Thutmose's generals encouraged him to take one of the easy routes. As he thought about it, however, he realized that the Prince of Kadesh and his troops would not expect them to come through the mountains because of its difficult terrain. They would likely be waiting at some point along the other two routes. So, to the disappointment of his generals, he decided to take the route through the mountains.[1]

And indeed, Thutmose was right. The Prince of Kadesh's men were waiting for them at the ends of the two easy routes. The prince had split his army into two groups, with one half of it in the south and the other half in the north. Furthermore, he had left almost no men to guard the city of Megiddo.

The following day Thutmose led his men through the treacherous pass, and when they broke out into the open, with the city of Megiddo directly ahead, they saw that it was lightly guarded. But Thutmose didn't want to attack the city at this point. He still had to defeat the prince's army. It was late in the evening, so he camped overnight and was ready for battle the next morning. He split his men into three wings and moved quickly to attack the flanks of both sections of the prince's army. They were so surprised by an attack coming from an unexpected direction that most of the men broke rank and fled. Most of them ran for the shelter of the city.[2]

Thutmose pursued them, and by the time he got to the city he could see that many of them were trapped. The defenders of the city had seen the fleeing men coming and had opened the gate, but as Thutmose's army came into view they immediately shut it, leaving many outside. The citizens inside, however, acted quickly; they lowered ropes made of clothes to pull the stranded soldiers over the walls.

Thutmose wanted to attack the city, but by now most of his soldiers were plundering the enemy camp, taking whatever they could find. By the time he got his army reorganized most of the enemy, including the Prince of Kadesh, were safe in the city, which had a high, strong wall all around it. Thutmose could see that it would be suicidal to attack it directly, so he decided on a siege. His troops had plenty of supplies, and there were more supplies available in the surrounding area. But the people within the city were cut off, so it was only a matter of time before they ran out of food and other supplies. The siege lasted for seven months, but finally the citizens and what was left of the army surrendered. By this time, however, the Prince of Kadesh had somehow escaped.

It had taken longer than he had hoped. Nevertheless, Thutmose had soundly defeated the prince's army, and he had captured Megiddo.

SUMMARY OF THE BOOK

Like all rulers or generals going to war, Thutmose III was looking for something that would give him an advantage, and he found it. In his case it was a tactic that gave him an element of surprise. Throughout history, and even today, military leaders contemplating war, or involved in it, are still looking for some sort of advantage over their enemy. Whereas Thutmose used a surprise tactic to his advantage, throughout most of history military leaders have searched for a new "wonder weapon"; in essence, a weapon the enemy does not have. As we'll see in this book, it is usually physics that provides a path to this new weapon. Physics and science in general has indeed been of tremendous value to military leaders. It has given them a better understanding of ballistics so that they can aim their guns better; it has given them radar so that they can detect the enemy before they are detected; it has given them an understanding of the electromagnetic spectrum so they can use radiation in various military applications; it has given them an understanding of rocketry and jet engines, and an understanding of the secrets deep within the atom so they are able to build super bombs.

This book gives an overview of most branches of physics, and it shows how they are used for military applications. It also gives a summary of the history of war all the way from the first bows and arrows and chariots through to the atomic and hydrogen bombs. We begin in chapter 2 with the Egyptians, Assyrians, and early Greeks. We'll look at some of their interesting weapons, such as the ballista, the onager, and the trebuchet, all of which involve basic principles of physics.

In chapter 4 we look at the rise and fall of the greatest military establishment ever seen up to that time, namely the Roman Empire. The early English-French battles are also included in this chapter; one of the most famous of these was the Battle of Agincourt, where the English used the longbow to overcome a much larger and more powerful army. It was their secret new weapon.

In chapter 5 we see the introduction of new technologies that completely changed the nature of war: gunpowder and cannons. Cannons were, in fact, so effective that they led to wars that lasted for a hundred years. At this stage, however, we can't say that physics made large contributions to the art of war because, for the most part, it didn't exist. But as we'll see in chapter 6, three men, including Galileo, made important advances and helped put physics on a much better footing.

With these advances and others, war became even more prevalent throughout Europe. Rifles improved significantly, beginning with the matchlock and ending with the flintlock a few years later. In addition, ships were now getting larger, and they were soon equipped with cannons. Furthermore, with William Gilbert's discoveries in relation to magnetism came a better understanding of navigation at sea, so sailors could now head out into the unknown without worrying about getting lost.

Then came the magnificent discoveries of Isaac Newton, and physics was raised to new heights of understanding. His discoveries are discussed in chapter 7. Following this came the Industrial Revolution, discussed in chapter 8. In a period of less than one hundred years the civilized world changed significantly. In particular, several new techniques, including mass production, made war even more devastating.

In chapter 9 we look at Napoleon and his weapons and tactics. Without doubt, he is one of history's greatest military tacticians, but strangely he didn't introduce many innovative new weapons. About this time another revolution in physics was occurring, and it would lead to a tremendous change in warfare. It began with the discovery that a "current" of electricity could be produced by a simple device called a pile. Soon the new phenomenon was all the rage throughout Europe, and it quickly attracted some of the best minds in physics:

Oersted, Ohm, Ampere, and Faraday. Electric generators, motors, and other electrical devices followed, and of course, they eventually became central to war.

In chapter 10 we come to the American Civil War, which was the most devastating war ever fought on American soil. By this time tremendous advances had occurred, including the percussion cap, which quickly led to much more accurate and deadly rifles, along with the first use of submarines, balloons, and the telegraph in warfare.

In chapter 12 we discuss the airplane. World War I erupted only a decade after the first flight of the Wright brothers. And it didn't take long before airplanes were used in the war. "Dogfights" were soon common, and the airplane has played a central role in warfare ever since. Many other new weapons were also developed in World War I. They included huge new cannons, the first tanks, poisonous gas, and flamethrowers.

Soon after World War I radar was developed, and it would eventually play a central role in war. Along with it came a significant improvement in submarines, and the use of sonar. Submarines would be very effective for the Germans in World War I and at the beginning of World War II.

Then in 1939 came another, even greater war, namely World War II, which produced phenomenal new weaponry. These developments included important advances in radar, the first jet airplanes, the first rockets, the first large computers, and of course, the atomic bomb. All of these will be discussed.

Finally, in the last chapter we will discuss the hydrogen bomb and some of the possible weapons of the future.

2

EARLY WARS AND THE BEGINNING OF PHYSICS

A s we will see throughout the book, every era had its "wonder weapon," and one of the earliest was the chariot. Pulled by two or three horses, chariots allowed warriors to move at tremendous speed. The chariots were usually manned by a driver and an archer equipped with a large number of arrows. Fast-moving chariots would crash into the infantry of the enemy while the archer fired arrows, frequently causing panic. Like our tanks of today, the chariot became a major weapon of early armies. Soon thousands of them were involved in the battles of the day.

BATTLE OF KADESH

One of the largest chariot battles the world had seen occurred in 1274 BCE near the village of Kadesh (in present-day Syria). More than five thousand chariots were involved. A large Egyptian army was led by twenty-five-year-old Ramses II. He was brash and confident but had little experience. Against him was a Hittite force led by Muwatallis II, who was a veteran of many wars and had considerable experience. Ramses led a force of about thirty-five thousand men, which included about two thousand chariots and a large number of archers. The Hittite army consisted of over twenty-seven thousand men and close to three thousand five hundred chariots. The Egyptian chariots accommodated two men, and they were much lighter, faster, and more maneuverable than the Hittite chariots, which were built to accommodate three men.

Ramses commanded four divisions, each of which was named for an Egyptian god: Amun, Re, Seth, and Ptah. He also had another division of mercenaries called the Ne'arin. With Ramses in the lead, his army began a month-long march toward Kadesh. When he was about seven miles from it, his men came upon two Bedouins, or nomads, who claimed they had been conscripted to serve in the Hittite army but had escaped. Ramses questioned them and was pleased

19

when they told him that Muwatallis's army was 135 miles away in a place called Aleppo. Furthermore, they said Muwatallis was afraid of Ramses and his army.

This made Ramses even more confident, since it meant he would be able to capture Kadesh without having to fight the Hittites. Without verifying the story he quickly pushed on. Indeed, he was in such a rush that he and his body-guards had soon outdistanced most of his troops. Close to his destination was the Orontes River, which was difficult to cross in most places, but it could be crossed close to Kadesh. Ramses and his small contingent of guards splashed their way across it, then moved through a wooded area to a clearing, from which he could see Kadesh. He decided to set up camp, and within a short time his Amun division caught up with him, but his other divisions were still relatively far behind.

As the men were setting up camp his guards brought two captured Hittite soldiers before him. Ramses began questioning them, but they refused to talk. Only after being beaten did they finally confess, and what they had to say shocked Ramses. They told him that the Hittites were massed behind the old city of Kadesh with infantry and chariots, and that they were more numerous than the grains of sand on the beach.

Ramses could hardly believe what he was hearing. The two Bedouins that he had talked to earlier had been lying, and indeed Muwatallis had sent them to set a trap. Ramses was now only a few miles from Kadesh, and he had only half his army with him. The Hittites were no doubt ready to attack. Ramses sent messengers to the lagging divisions telling them to hurry. He knew that the Ptah division was not far away, however, and with it he would have three-quarters of his army, so he wasn't worried.

Muwatallis, meanwhile, had divided his troops into two main forces. One was to strike at the rear of the Egyptian army; the other, which included Muwatallis, himself, along with a force of one thousand chariots and a large contingent of infantry, would strike from the side, preventing the Egyptians from retreating.

The Hittite chariots spread into formation, then they attacked. The Re division, straggling behind, had just emerged from the forest into the clearing area. Twenty-five hundred Hittite chariots ripped into it; the Egyptians didn't know what hit them. Panic raced through the survivors as the Hittites slaughtered most of them. The remnants of the division ran toward the safety of the main Egyptian encampment, but the Hittites followed. Ramses was surrounded by his guard, which consisted of the best-trained troops in his army. The Hittite charioteers rushing toward him proved no match for the well-trained guards, who quickly killed large numbers of them.

Ramses had been busy reaming out his officers when the attack came, but he quickly took charge, and with what remained of his army he counterattacked. He did, however, have several advantages: his chariots were faster and could easily outmaneuver the Hittite chariots. Furthermore, his archers had a relatively powerful composite bow, and within a short time they had inflicted severe damage on the Hittite forces.

Strangely, the Hittite infantrymen, who were sure the battle was almost over, stopped and began looting the Egyptian camp. As a result, they became easy targets for the Egyptian counterattack. They were soon routed, with many of them dead on the field. The battle, which had started out as a slaughter for Muwatallis II, was now turning in favor of the Egyptians. Nevertheless, Muwatallis ordered another attack. In the meantime, Ramses' Ne'arin troops arrived, bringing his army to nearly full strength, and it counterattacked with everything it had. Soon the Hittites were overwhelmed, and many of them fled back toward Kadesh.

But Muwatallis II was not ready to give up. Most of his chariots, however, were now on the opposite side of the Orontes River. They had to cross it to attack the Egyptians. Ramses looked over the situation and decided to let them attack; he had a plan. He let the Hittite chariots cross the river, knowing that as they started up the steep bank toward the Egyptians they would slow to a crawl. When they did, Ramses ordered his chariots to attack, and they soon pushed the Hittites back into the water, inflicting heavy losses.

Muwatallis then ordered another charge, and again his troops were driven back, this time with even heavier losses. For the next three hours, in fact, Muwatallis continued the same tactic, until most of his officers were gone and many of his charioteers had been killed, many by drowning. Finally, when the Ptah, the last of the Egyptian divisions arrived, Muwatallis decided it was hopeless. He and his troops retreated, many to the safety of Kadesh, with others continuing on to Aleppo.

Ramses had also lost a lot of men by this time. He decided not to attack Kadesh and instead returned to Egypt. Both leaders claimed they had won the battle, and, indeed, Ramses had routed the Hittites, but he had not achieved his goal, namely the capture of Kadesh. Muwatallis, on the other hand, claimed he had stopped the Egyptians, and indeed, they had left.[1]

THE WONDER WEAPON

The chariot obviously played a large role in the Battle of Kadesh, and for many years thereafter it continued to be a major weapon of war. And certainly when it was first introduced it created terror among enemy troops. Most of the first chariots were built for two men, but later three- and even four-man chariots were used.

Most people are familiar with chariots from the movie *Ben Hur*, which starred Charlton Heston. It contained an exciting nine-minute chariot race that became one of the most popular sequences in the history of film, and it certainly gave viewers a good idea what it was like to drive or ride in a chariot.

An early war chariot.

Although the chariot was initially a wonder weapon, it didn't take long before many armies had them. So of course a search soon began for a new wonder weapon. At the time, weapon designers couldn't turn to science because science didn't yet exist. Nevertheless, the search was on for a new weapon that would shock and terrorize the enemy. Indeed, the process soon became an endless cycle.

COPPER, BRONZE, AND IRON

Wonder weapons had, indeed, already appeared. The earliest weapons were no doubt wooden spears and sharpened knives made of stone, but about 5000 BCE men in Persia and Afghanistan began to find strange lumps that could be hammered into various shapes, and they soon discovered that the material could be melted at a relatively modest temperature. It was what we now know as copper, and it would soon play a large role in the lives of the people of the time. Copper could be molded or cast into many different shapes. But it was soft, so knives made of copper would not keep a sharp age. Something harder was needed, and perhaps by luck, or perhaps by extensive experimentation, it was discovered that when a softer metal, tin, was added to copper in a molten state, the result was a new metal, bronze, which was considerably harder than either copper or tin alone. Bronze was soon used for knives, spears, and other weapons that needed sharp edges.[2]

The science of metals, or metallurgy as it came to be called, soon came into its own. Axes, daggers, shields, and even helmets were cast from bronze, and they soon became the new wonder weapons of war. For years however, people had been aware of a red-brown mineral that could be found near the earth's surface, and it was eventually discovered that it could be mined and smelted into another new metal, iron. It's hard to say exactly when the Iron Age began. Iron had been observed as early as 3000 BCE, but it took until about 1200 BCE before suitable smelting techniques developed. Iron smelting is much more difficult than copper smelting because of iron's higher melting point. Furthermore, when it was first obtained in a relatively pure form, it wasn't much harder than bronze, but then it was discovered that if carbon was added to it, it became much harder.

One of the things that may have spurred the search for a better metal than bronze was that tin was relatively rare, and shortages frequently occurred. Another reason was that states that could not afford to build thousands of chariots needed weapons that could match the dominance of the chariot. Infantry was no match for chariots, but some leaders began to believe that with the proper weapons it could be. And as metallurgists learned more and more about iron, and how it could be strengthened with carbon, much longer swords and spears came into existence along with iron shields that arrows could not penetrate. And with them, infantrymen could be equipped to stand up to chariots. With shields that could easily deflect arrows, and iron helmets to protect their heads, they could attack chariots with long iron swords and spears.

THE ASSYRIANS

The chariot was still a lethal weapon for many years, but a real breakthrough came when warriors on horses began to challenge them. Among the major enemies of the Assyrians were the nomads and barbarians of the north countries. Their life centered on horses, and they were particularly comfortable around them, usually learning to ride at a very early age. And it soon became obvious to them that a man on horseback, equipped with a bow or sword, had an advantage against chariots, since the mounted warrior was very mobile and could easily outmaneuver a chariot. He was high off the ground, and, with a horse under him, he was a formidable force; in addition, he was fast in a charge, even faster than a chariot.[3]

We now refer to forces of men mounted on horses as cavalry, but at that time they were not organized into what we would normally think of as cavalry. Nevertheless, they were effective. Early horse-mounted warriors didn't use saddles, but they were quite comfortable and stable without them. Stirrups came even later.

Mounted warriors from the north were highly effective in their frequent attacks on the Assyrians. It wasn't long, therefore, before the Assyrians began to develop their own cavalry. The Assyrians, who eventually became the most powerful empire of the region, were descended from the Akkadian Empire, which flourished near the upper Tigris River (near present-day Iraq) and lasted until about 2100 BCE. The Akkadian Empire eventually evolved into two states: the Assyrians in the north, and later the Babylonians in the south, but it was the Assyrians who first came to dominate the region.

During the early Assyrian years, the Bronze Age was in full swing and most weapons were made of bronze. Over the years the regional power of the Assyrians fluctuated, but there were two eras in which they were particularly powerful. Their early period of power and empire lasted from 1365 BCE to 1076 BCE. During this time their armies conquered most of the surrounding countries, including Egypt, Babylonia, Persia, Phoenicia, Arabia, and Israel. But after 1076 BCE Assyrian dominance ebbed. Then, in 911 BCE, the Assyrians once more began to grow in power. The Assyrian Empire eventually became the greatest military power the world had ever seen up to that time. Its resurgence was mainly due to Tiglath-pileser III, who ascended to the throne in 745 BCE.

Tiglath-pileser III began by introducing dramatic changes. First he increased the efficiency of the Assyrian administration. Then he turned his attention to the army, which had become significantly weakened over the years. At this time the only army that existed was quite small; when a large army was needed, recruiters were sent out to round up farmers and anyone else they could get, and the conscripts were usually given very little training. Tiglath-pileser set up a large standing army, one of the first in history. And the soldiers were given uniforms and some of the best weapons of the time. He also significantly improved the roads throughout Assyria.

An Assyrian warrior.

Chariots were still being used, but Tiglath-pileser immediately saw the advantage of cavalry, setting up a large cavalry division. The Assyrians did not have much experience with horses, and they were initially not nearly as good on them as the barbarians were. But with training they improved. At first the Assyrian cavalrymen worked in pairs, with one controlling the horses and the other shooting arrows. But soon each warrior had his own lance and control of his own horse. Cavalry eventually became the core of the Assyrian army, with thousands of cavalrymen on horses. This meant, of course, that large numbers of horses were needed, and Tiglath-pileser also took care of this. Large stables were set up to raise and care for horses.

There's no doubt that the Assyrians were a "warring nation" right from the beginning. They were, in fact, at war most of the time they were in power. And under Tiglath-pileser they continued their warring ways, conquering country after country. Not only did Tiglath-pileser build up the cavalry; he also significantly improved the infantry. The infantry consisted of archers, shield bearers, slingers, and spearmen. Slingers, who threw stones, were frequently used to distract the enemy. Large shields were used by most nations to protect their forces against the onslaught of arrows. Arrows were usually fired high so they would drop down on the enemy; the shield bearers would therefore hold their shields over their heads to protect the infantry. Tiglath-pileser employed slingers to project stones directly at the enemy, and to protect themselves they had to lower their shields. The Assyrian archers would then fire over their heads so that falling arrows would not be deflected by their shields. Tiglath-pileser also introduced lancers; they were soldiers with particularly long spears, called lances. They were much longer than swords are daggers, and, as a result, when they were used in an attack, swords were quite ineffective against them.[4]

There was, however, a serious problem for the Assyrians. So many nations were at war at this time that cities and towns were in constant danger of being invaded, not only by other nations, but even by their neighbors. And they needed protection. Kings and rulers, with their huge egos and aggressive ways, were always hungrily eyeing the resources and wealth of their neighbors and neighboring countries. Few were satisfied with what they had. War was a natural thing, and they had to go to war not only to conquer new lands, but also to build up their treasury.

The Assyrians were certainly guilty of this. Furthermore, it was well-known to all their enemies in the surrounding countries that they were brutal. They frequently killed entire populations, and they killed indiscriminately. They also used mass deportation as a terror weapon. If there were uprisings in any of the countries they conquered, they would deport thousands of people to other lands.

Tiglath-pileser was well known for this. For example, in 744 BCE he deported sixty-five thousand people from Iran to the Assyrian-Babylonian border, and in 742 BCE he deported thirty thousand people from Syria to the Zagros Mountains in present-day Iran.

Because of these practices, many people put considerable effort into building huge walls around their towns or cities for protection. These walls were usually several feet thick and at least twenty feet high. Several years were frequently spent building them. The earliest walls were made of mud mixed with various other materials; they were thick enough to give some protection, but it soon became obvious that they were vulnerable. Mud was not very strong. Nevertheless, an enemy would frequently bypass a city if its walls were too thick and high. It was usually too much trouble for them.

Walls, however, were merely a challenge to the Assyrians. They weren't going to let walls stop them, and they soon began designing and building siege engines to get through them. Actually, they were nothing more than huge battering rams constructed of wood. In many ways they resembled a gigantic tank on wheels. They usually had four wheels, but some of the later ones had six. And because they were so big and heavy, it usually took thousands of soldiers to move them.

As terrifying as they were, the defenders usually fought back with all they had. The siege engines would have to be pushed up to the edge of the wall, and it soon became obvious that the pushers and anyone inside would need protection while it was being moved forward, since the defenders would bombard them with arrows and rocks, and as the siege engine moved close to the wall they would try to set it on fire. For protection, the Assyrians built small towers atop their siege engines for their archers. These archers would fire back at the defenders as the siege engine was being moved forward.

When the siege tower, which was frequently several stories high, reached the wall, a huge battering ram with an iron (or bronze) "bit" was rammed again and again against the wall. It was powered by a large contingent of soldiers. Slowly it would chip away at the wall, and as this took place, fierce fighting would go on between the Assyrians and the defenders. Fire, of course, was a major weapon of the defenders, so the Assyrians had to cover their engine with a huge sheet of animal skins that was kept wet.

As time passed, walls were built thicker and thicker, and eventually stone walls were used. But the Assyrians merely built bigger and bigger siege engines with more effective metal bits. As stone increasingly came to be used for the construction of city walls, it became more and more difficult for the siege engines to batter them down. Nonetheless, they continued to have some success. One of the largest siege engine of ancient times was the Greek *helepolis*; over one hundred feet high and so stable it could not be tipped over, it far outstripped the scale of Assyrian siege weapons.

Over time, the Assyrian Empire began to weaken. It had collapsed by about 610 BCE.

GREEKS AND THE BEGINNING OF PHYSICS

While the Assyrian Empire began to fade, other nations began to flourish, including Babylonia, the Persian Empire, which lasted to 330 BCE, and Phoenicia, the seafaring state that lasted to about 539 BCE. But the ancient civilization that had the biggest influence on physics was Greece, which consisted of city-states that began coming into power about 800 BCE. Indeed, before the Greeks, there was little if anything that could be called physics, and there was little science in general. Furthermore, the first scientists were not referred to as such; they were referred to as philosophers. But there's no doubt that one of their major aims was to understand the world around them. They were particularly interested in motion and matter. Why did things fall? And what exactly was the role of air, water, fire, and the earth beneath their feet? What was time? And their curiosity extended to the sun, the moon, and the stars. How far away were they? How big were they? Why did they seem to move?

The first science was, no doubt, a form of physics. It was not what we think of as physics today, but it did include many of the same topics. It was drawn from astronomy, mechanics, optics, and areas of mathematics such as geometry. The early Greek philosophers set out to understand the mysteries of the earth and the known universe, and although they arrived at some ideas that may seem strange to us today, they did make important advances. One of the biggest advances was to move away from mythological explanations for the phenomena they observed. Instead, they developed logic and learned to look for reasonable and logical explanations.

One of the first of these philosophers was Thales, who lived from 624 to 546 BCE. He was the first to emphasize the importance of explanations based on reason, and he was particularly interested in why things happened. Because of his contributions he is sometimes referred to as the father of science. He is said to have predicted the eclipse of May 28, 585 BCE. There is some controversy about this, however, as most modern astronomers feel that such a prediction was not possible at that time. But there is no controversy about his most important contribution. At that time Greek sailors never left the sight of land because they had no idea how to navigate when no land was visible. Thales showed them how to use Polaris (the North Star) for navigation. He also studied the strange phenomena associated with magnetism and amber, and he took a serious interest in the phenomena of time and the basic nature of matter.

The two major philosophers who came after Thales, namely Socrates and Plato, were both giants of rational thought, but their interest was mostly in logic, philosophy, and mathematics. Socrates was considered one of the wisest people of his time, but science was not central to his thinking. Plato, a student of Socrates, was probably most famous for his founding of the Academy of Athens.

In 384 BCE, however, the ancient philosopher best known to us was born: Aristotle. He was highly influential in his own era, and he remains influential today. He was strongly interested in science, and he made several contributions, but because his influence has extended over such a long period of time, he is frequently regarded as someone who hindered the development of science. Nevertheless, his goals were admirable. As he stated in his writings, his main aim was to discover principles and causes of change, and not just describe them. Much of what he came up with, however, was erroneous. One of his major hypotheses was that there were four basic elements: earth, water, air, and fire. And he postulated that everything was made up in some way from these four elements. He also had a strong interest in the phenomena of motion, and he classified all motions as either "natural" or "violent." A falling object had natural motion; a thrown object had violent motion. He also believed that everything beyond the earth—sun, moon, and stars—was made up of a fifth element he called "ether."[5]

A number of other Greek scientists of the time also made important contributions. Eratosthenes (276–194 BCE) invented a system of latitude and longitude for the earth. He also calculated the circumference of the earth using the shadows of sticks at different positions. In particular he pointed out that if the earth was flat, there should not be shadows from vertical sticks at different positions at the same time (only one could be shadowless) when the sun was directly overhead. He used his new knowledge to calculate the circumference of the earth to be two hundred fifty thousand stadia (we're still not sure however what a stade is). He also calculated the distances of the sun and the moon, giving us first, but very approximate, estimates.

Another important early Greek scientist was Hipparchus, who was born in 175 BCE. He gave us more accurate measurements of the distances to the sun and to the moon, and he was the first to set up a catalog of most of the visible stars.

Physics first appeared as a result of the studies and speculations of the above philosophers. It's important to note, however, that their contributions came almost entirely from "thought." Experimental physics was not known at the time, and, indeed, the early philosophers did not perform any experiments in their attempts to prove their ideas. Nevertheless, even at that time they realized that there was a difference between what we call "pure physics" and "applied physics." Pure physics is usually thought of as the accumulation of knowledge about the physical aspects of the world and universe, such as the basic principles of space, time, matter, motion, and so on, with no thought of how this knowledge should be applied. Applied physics, on the other hand, is the application of this knowledge to assist society in some way. At that time, the main application of physics was the design and manufacture of war machines. Early philosophers such as Socrates, Plato, and Aristotle argued that science should not necessarily have applied goals, particularly applications to war. Knowledge should be accumulated for its own sake.

In spite of the arguments against doing so, it wasn't long before the new discoveries of physics were being used to build new weapons of war. Many of the early advanced weapons built by the Greeks were based on an important physics concept called torsion. In physics, torsion is the twisting of an object due to an applied torque, where torque is a twisting force. And indeed, torsion soon became the basis of new terror weapons that were frequently referred to as machines or engines.

THE NEW WONDER MACHINES

The most common new wonder machines that came out of Greek physics (though not necessarily constructed by the Greeks) were the ballista, the onager, the trebuchet, and various other types of catapults. We talked about siege machines earlier that were used to break through walls; some of the above were also eventually used as siege machines. Let's look at each of them in turn. The ballista was invented by the Greeks and later modified and used extensively by the Romans. It was similar to a giant crossbow, but it used torsional energy that was stored in twisted skeins. Two wooden arms were used to twist the skeins; ropes were attached to one end of each of them, with the ropes extending back to a "pocket" that held the projectile. The ropes were pulled back by a winch. It had a trigger on it, and when everything was ready to go, the trigger was pulled. Various types of projectiles were used, including stones, darts, shaped poles, and even body parts. It could throw them several hundred yards.[6]

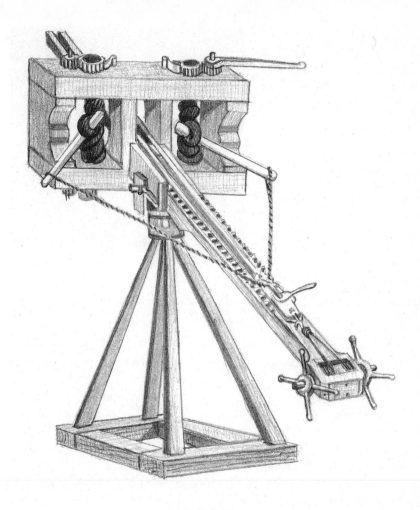

The ballista.

A variation on the ballista came a little later in the form of the onager, which was used mainly by the Romans. It also used torsion, but was basically a type of catapult. It consisted of a large frame that was placed on the ground. A vertical frame of wood was rigidly attached to it. This vertical frame contained an axis that had a spoke or arm projecting out of it. This spoke was attached to ropes (or a spring) that could be twisted; the arm was pulled back, or armed, against the buildup of torsion in the ropes. Again, there was a pin to release it, and when the pin was struck with a hammer the projectile was launched toward its target. Large stones were usually used as the projectiles.

Soldiers arming an onager.

The third type of new weapon, the trebuchet, was actually the most powerful of the three. It was invented by the Romans and had three main characteristics:

- It was not powered by torsion. Its power came from gravity acting on a counterweight.
- It used what is called the "fulcrum principle," where one arm was much longer than the other. The throwing arm was usually four to six times longer than the counterweight arm.
- A sling with a pouch was attached to the end of the throwing arm to increase the speed of the projectile.

The device was loaded by placing a large and usually very heavy stone in the pouch. The throwing arm was then pulled down against the weight of the counterweight. It was tied down until ready. When it was triggered, it could throw rocks of three hundred pounds and more a distance of several hundred yards, but it was not nearly as accurate as the ballista or onager.[7]

The trebuchet and the onager were both a form of catapult. A catapult is a device that usually has an arm that is pulled back against a force and released. Several other types of catapults were also used, but the major ones were the two above. The physics of the above devices will be discussed in the next chapter.

ALEXANDER THE GREAT

One person who made extensive use of the new weapons was Alexander the Great. Born in Pella, the capital of Macedonia in 356 BCE, Alexander became the greatest military leader of his time, conquering most of the known world. Taught by Aristotle beginning at the age of sixteen, he developed a strong interest in science and physics. When he turned nineteen he began accompanying his father, Philip II, on some of his campaigns. Shortly thereafter, however, his father was assassinated, and because his father had multiple wives, and Alexander's mother was only one of them, his chances of inheriting the throne were not good. But he was determined to get it, and he took the necessary steps, killing several people in the process.[8]

When he became leader he quickly undertook a series of campaigns that lasted for almost ten years. In the end he had conquered Egypt, Mesopotamia, Persia, Central Asia, and even India. And by the time he was thirty he was considered one of the greatest military leaders the world had ever produced.

Aristotle had instilled in him a love of knowledge, and it remained even after he became king. As a result, he created one of the greatest learning centers the world had ever seen. After conquering Egypt, he founded Alexandria in 331 BCE, setting it up as a scientific research center. Although he only stayed in the city for a few days, he left with his viceroy and a general named Ptolemy an outline of the work he wanted done.[9] At Alexandria what was called a "mouseion" was set up for the study of engineering, astronomy, navigation, physics, and the machines of war. The best scientists in the country and in surrounding countries were then invited to study there, including Eratosthenes and Hipparchus.

Perhaps the greatest feature of the new Mouseion at Alexandria was its library. It eventually became the largest library in the world, housing over seven hundred thousand manuscripts. It thrived for centuries, but much of it was eventually destroyed by fire.

ARCHIMEDES

One of the men who studied at Alexandria was Archimedes, who was born into 87 BCE in Syracuse, Sicily. He made a large number of contributions to physics, one of the most important of which was a principle now referred to as Archimedes' principle. It states that *a body immersed in a fluid experiences a buoyant force that is equal to the weight of the fluid it displaces*. He also designed what is now called Archimedes' screw. According to early accounts, the king of Syracuse commissioned Archimedes to design a large ship, but it was soon discovered that a considerable amount of water was leaking into the hull, and it was difficult to bail it out. Archimedes designed a machine with a revolving screw-shaped blade inside a cylinder that raised the water from the bottom of the hull as it was turned.[10]

Archimedes was also one of the first to explain the principle of the lever. And he is reported to have helped the people of Syracuse when they were attacked in 14 BCE. Presumably, he set up large curved mirrors that reflected the rays of the sun upon the attacking ships, causing them to catch fire. Most modern scientists doubt that this was possible at the time.

3

BASIC PHYSICS OF EARLY WEAPONS

Physics is related to the early weapons of war just as it is to the more sophisticated later weapons. So far we have talked mostly about chariots, men on horses, bows and arrows, spears, and such things as the ballista, the onager, catapults, and trebuchets. Physics is involved in all of these things, but we haven't shown how it is involved. In this chapter we will do this, but first we will discuss the basic concepts of physics, beginning with the most basic ones, such as speed and acceleration, and proceeding through to more complicated ones, such as energy and momentum.

VELOCITY AND ACCELERATION

Everyone knows that if you shoot an arrow into the air it rises to a certain point before falling back to earth. It's also known that its speed as it leaves the bow depends on how hard the string pushes on it, and it's easy to see that its speed throughout its flight is not the same. After all, if you shoot it straight upward, it stops at some point before it starts to fall back to earth.

We have a slight problem in relation to motion on earth, however. Every object that moves has to pass through air, and this air has an effect on its speed as well as the path it takes. Dealing with the effect of air, however, is rather complicated, so for now we will ignore it.

The first thing we can say about an object in motion is that it has a certain speed relative to the surface of the earth. Speed is a useful concept, but even better (as far as physics is concerned) is *velocity*. Speed is defined as the distance something travels in a unit of time, say, a second, or even in hour. An arrow, for example, can have a speed of fifty feet per second. The problem with this is that it doesn't tell us anything about the direction that the arrow is traveling. If we specify both speed and direction, we have velocity. The velocity of the above arrow, for example, might be fifty feet per second in a northern direction.

If we look at this arrow a little closer, however, it's easy to see that it doesn't have a constant velocity. Its velocity is continually changing, and the biggest change will occur when it is shot directly upward. After all, it stops at its highest point. We refer to this change in velocity as *acceleration*. The arrow might leave the bow with a velocity of fifty feet per second, but a few seconds later it will be going only ten feet per second. Acceleration is clearly different than velocity, and it therefore needs a different unit. The unit in this case is feet per second squared (in the metric system it is meters per second squared). Velocity and acceleration are related by a simple formula: velocity (v) equals acceleration (a) × time (t), or more simply $v = at$.[1]

FORCE AND INERTIA

Closely related to velocity and acceleration is another important physics concept called *force*. For an arrow to gain speed—in other words, to accelerate—it must undergo a force, and as I mentioned earlier, it is the string of the bow that applies the force to the arrow. A force is simply defined as a push or a pull. And force is like velocity in that it has both magnitude and direction (we refer to such a quantity as a vector).[2]

We can relate force to acceleration, but before we do, let's introduce another important concept from physics. Everyone knows about weight, and how it seems to creep up on you when you eat too many chocolates. What we're interested in is closely related to weight, but it's not exactly the same. We refer to it as *mass*, and we abbreviate it as m. The mass of an object is its weight divided by the acceleration of gravity, which is usually abbreviated as g. I'll explain a little later why we need mass rather than weight.

The relationship between force and acceleration was given by the English physicist Isaac Newton. He included it in three laws of motion that he published in his *Principia* in 1687. He explained that an acceleration created by a force acting on a body is directly proportional to the magnitude of the force and inversely proportional to the mass of the body. We can write this an algebraic form as $a = F/m$. As it turns out, it is more convenient to use metric units in this formula (instead of the units you are probably more familiar with, namely feet, miles, and so on, which are units in what is called the British System). Within the metric system, however, we have two systems, referred to as *cgs* (centimeter, gram, second) and *mks* (meter, kilogram, second). In the mks system, acceleration is measured in meters per second squared, mass is measured in kilograms, and the unit of force is the Newton. In the cgs system, acceleration is measured in centimeters per second squared, mass is measured in grams, and the unit of force is the dyne, which is the force required to cause a mass of one gram to accelerate at a rate of one centimeter per second squared.

The above formula is usually written as F = ma. So the force on an object is the product of mass and acceleration. For example, if you want to create an acceleration of 25 km/sec^2 in an arrow with a mass of .01 kg you would need a force of .01 × 25 = .25 Newtons.

Closely related to the concept of a force is what is called *inertia*. We encounter inertia every day; when you push on an object or lift it, you have to exert a force to get it going. If an object is not moving—in other words, it's just sitting there—it tends to resist motion, and it takes a force to get it going. Indeed, the heavier it is, the greater the force that is needed. This "resistance" to a change in motion is called inertia, and Newton described it in his first law of motion: *a body will continue in a state of rest or uniform motion in a straight line unless acted upon by a force*. Note that this applies not only to something at rest, but also an object in uniform motion.

This means we need a force to overcome inertia, and this force produces acceleration according to the above formula. Furthermore, a force is always associated with two bodies. If one body is being pushed, the other body has to do the pushing. This also applies to an object sitting on the floor; because of its weight it pushes down on the floor. But according to Newton, the floor pushes back with an equal force in the opposite direction. Newton postulated this in his third law of motion: *whenever a body exerts a force on a second body, the second body exerts a force on the first body*. These forces are equal in magnitude and opposite in direction. They are frequently called the "action" and "reaction" forces. A good example of these forces can be seen when you hold a garden hose with water pouring out of it. You feel a backward force on your hands; this is a reaction force, and it's why rockets work: explosive gases shoot out the back of the rocket, giving the rocket its forward thrust.

MOMENTUM AND IMPULSE

Another important concept in physics is *momentum*; it is the product of mass and velocity (m × v). It is particularly important when one object collides with another. As you no doubt know, when a massive object collides with a smaller, lighter one, it's the smaller one that suffers the most. To understand this more fully we have to introduce the concept of *impulse*. Assume a soldier hits the shield of another soldier with his sword; he's obviously applying a force to it, but this force is only exerted over a short period of time. The product of this force, and the time it acts, is defined as impulse. Furthermore, it's obvious that this impulse is going to cause the shield to move with a certain velocity, and this velocity will depend on the mass of the shield. So impulse is somehow also related to momentum. Indeed, the impulse creates momentum; or, to be more precise, since the momentum of the shield was zero before the impulse, the impulse creates a *change* in momentum. So impulse is equal to change in momentum.[3]

Now let's go back to our discussion of the collision of two bodies. Of particular importance in relation to such a collision is what is called *the principle of conservation of momentum*. It states that the total momentum of any isolated system remains constant. This means that the total momentum before the collision will be equal to the total momentum after it, assuming there are no outside influences. Let's assume that the collision is a head-on collision, and that both objects have the same (but opposite) momentum. It's pretty obvious that they will stop dead. It almost seems as if their momentum has disappeared—but it hasn't. Before the collision they had equal but opposite momenta, and the sum of two equal and opposite numbers was zero. After the collision it's still zero. As the collision occurred, each object imparted an impulse on the other, but the impulses were equal and opposite, so the objects stopped.

It's easy to see from this that if one of the objects has a greater momentum than the other, it will generate a greater impulse on the second, and if the two objects stuck together when they collided they would continue with a certain velocity in the direction of the one that had the greatest momentum.

THE EFFECT OF GRAVITY

Everyone knows that when you shoot an arrow at some angle in an upward direction, it doesn't travel in a straight path. It travels upward for a while then heads back toward earth, eventually landing. This is because of the gravitational pull of the earth on the arrow. In reality, the two objects are attracted to one another, but because the earth is so much more massive than the arrow, it appears to us that the earth is attracting the arrow. Again, it was Newton who explained what is going on. He postulated that all objects in the universe attract one another. Indeed, he even gave a formula for the force between any two objects.

Let's begin by considering a stone held at some distance above the ground. It is attracted by the earth with a certain force, and if we let it go, it accelerates downward until it strikes the ground. With a relatively simple apparatus we can measure this acceleration, and we find it to be 32 ft/sec^2, or in metric units, 9.8 m/sec^2.

Gravity is particularly important in relation to warfare because all objects, such as arrows, cannonballs, bullets, and so on, are affected by it. Such projectiles trace out trajectories that depend on several things, such as their mass and speed, and also the air pressure. (We'll talk about trajectories in detail later in the book.)

The acceleration of gravity is not the same everywhere, however. It depends on the mass of the planet you happen to live on. So if you were to travel to Mars or Jupiter, it would be different. As a result, your weight would also be different. On Jupiter for example, you would weigh 2.34 times your weight on earth. What is constant is your mass; it doesn't depend on the gravitational field you happen to be in, and that's why it is used in most of the basic physics equations. The relationship between mass (m) and weight (W) is given by $W = mg$, where g is the acceleration of gravity.

ENERGY AND POWER

If you lift something in an upward direction through a certain distance you do work. To perform this work it takes *energy*, and, as it turns out, there are several different forms of energy. Two of the most common forms are energy associated with motion, and energy associated with position. Energy associated with motion is called kinetic energy, and since it depends on motion, it will also have to depend on velocity. Furthermore, an object with a greater mass will have more kinetic energy than one with less mass, so kinetic energy also depends on mass. It is therefore defined as kinetic energy $= 1/2mv^2$, where m is mass and v is velocity. Its units are foot-pounds in the British System and Newton-meters in mks units.

Energy of position is called potential energy. It also has the ability to do work. Consider a stone held at some position above the ground. If you drop it, it does work on that dirt it strikes; it compresses and heats it slightly. We define it as potential energy $= mgh$, where m is mass, g is acceleration of gravity, and h is the distance above the ground from which it was dropped.[4]

Like momentum, energy is also conserved. In short, the conservation of energy states that energy can neither be created nor destroyed; it can only change from one type of energy to another. This can be demonstrated quite nicely if you take a ball and throw it upward. When first thrown, the ball has a high velocity, and its energy is therefore mostly kinetic energy. As it continues to rise, however, it gradually decelerates because of the pull of gravity. Finally, it stops, and at this point it has zero velocity, and therefore it has no kinetic energy. In essence, all its kinetic energy has been converted to potential energy, so at this point it only has potential energy. As it begins to fall again, however, its speed increases, and its kinetic energy also increases. At the same time, its potential energy decreases, and by the time it is just about to hit the ground all its potential energy has been converted back to kinetic energy.

The two above types of energy are not the only two types of energy. Other forms are deformational energy, heat energy, sound energy, electrical energy, chemical energy, and nuclear energy. You might, for example, ask what happens to the kinetic energy of the ball when it hits the ground. It appears to be lost, but it isn't. It is converted to deformational and heat energy.

In many cases we are not interested in the work (or energy) that has been done, but rather the rate of the work, or the amount of work accomplished per unit of time. This is referred to as *power*. In mks units, power is measured as joules per second, and by definition a joule/sec is a watt.

ANGULAR MOMENTUM AND TORQUE

Another type of motion that is important in relation to warfare and weapons is rotational motion. The wheel, or anything that spins about an axis, has angular or rotational motion. And just as we have linear velocity and linear acceleration, we also have angular velocity and angular acceleration. Angular velocity is measured as the number of revolutions per unit of time. Another common unit is the number of radians (rad) per unit of time, where the radian is $360/2\pi \approx 57.3$ degrees (360 is the number of degrees in a circle and π is the circumference of a circle divided by its diameter, which is 3.1416). Angular speed can, of course vary, and when it does it becomes angular acceleration. Its units are revs/sec^2.[5]

In the same way we have a concept analogous to force. It's the force that causes the rotational motion, and it is called *torque*. It has to be applied at some distance from the rotational axis to cause rotation, so there is also a distance involved. Torque is defined as force × distance (f × r). Note that you apply torque every time you use a wrench or open a door.

Earlier, in the case of translational or linear motion, we also had momentum, and in the same way we have angular momentum in this case. To determine the formula for it we must replace mass (m) and velocity (v) by appropriate angular quantities. Velocity is no problem; we merely replace it with angular velocity (ω), but m is a little more complicated because we are dealing with a large number of small masses, each at a different distance from the axis. If we add up all the little contributions from these small masses we can determine what is called the moment of inertia; it is designated by I. Angular momentum is then Iω.

MACHINES

Many of the early weapons were what we call machines in physics. A machine is a device that makes work easier. A simple example of a machine is a long board used to raise a box that is too heavy for us to lift. If you place one end of the board under the box, and place a block (called a fulcrum) a few feet away, then apply a downward force to the other end of the board, you find you can easily raise the box. This makes sense because work is force × distance; when we apply the force the box is raised a smaller distance than the top of the board moves. We are, in effect, using the extra distance to get a greater force. The work done is equal, but it is easier for us because we only have to apply a fraction of the force we would have to if we were lifting the box directly. This is basic to all machines.[6]

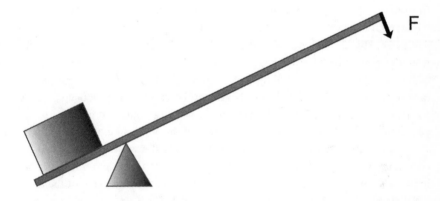

Many types of machines exist, and the principles associated with each of them were used in various early weapons. Some of the more common machines are:

Pulleys: They allow heavy loads to be lifted with less force; you merely have to move the rope a greater distance than the load moves.

Wheel and axle: A longer twist at the outer edge of the wheel exerts a more powerful but shorter motion near the axis.

Screw: Applying a larger but easier rotary force creates a smaller forward motion.

PHYSICS OF THE BOW AND ARROW

The bow and arrow was used extensively in early warfare. Archers were trained from an early age. In some cases they advanced toward the enemy on foot, carrying a shield; in others cases they rode in chariots. As we saw in the last chapter, chariots usually had a driver and an archer, and when the chariot got close enough to the enemy, the archer would begin firing arrows as fast as he could.

A bow, in essence, is a simple machine that changes one type of energy into another, making it easier for the archer to give the arrow a high velocity. What is needed for a high velocity is a rapid and forceful arrow movement, and of course muscles can do both of these, but not at the same time. To understand the physics of the bow and arrow let's begin with the archer loading an arrow and pulling the string back slowly. He is using his arm muscles to do this. He pulls the string back to its maximum extension, and in the process the bow bends. The energy from the archer's muscle contraction is stored in the bending of the bow. This is potential energy. He then lets go of the bowstring, at which point the string moves rapidly to the normal rest position. In the process it transfers energy from the bow to the arrow. In essence, potential energy is transferred to kinetic energy, as in the case of a falling ball. The transfer is obviously very rapid, and this gives the arrow a high speed. Note that the archer has produced a certain amount of energy, and by the conservation of energy, this energy must remain constant. But the bow can move with both a high force and a high velocity in a way that his arm cannot. The bow is a machine that stores energy. Muscle power is used to load the machine at low speed, then the machine releases the energy at high speed. Indeed, if you know all the variables, such as the mass of the arrow, the distance the bow is pulled back, and what force it exerts, you can equate potential energy to kinetic energy and determine how fast the arrow will leave the bow. Furthermore, if you know the angle at which is it is aimed (and ignore air pressure) you can determine how far it will go.[7]

Over the years bow and arrows were gradually improved. Several factors are important in relation to how powerful a bow will be. Three of the most important are its length, its shape, and its composition. In general, the longer the bow, the more powerful it will be, but other factors play a large role. We will see later that the English developed a very effective longbow and used it with considerable success against the French.

The overall shape of the bow is also important. Early bows had a single curve and were made of wood. Eventually, however, archers determined that if the ends of the bow were curved away from the user, the arrow would go farther. This was because the curving shortened the distance between the bow and the string at rest, and as a result, the string traveled farther before coming to a stop as it released the arrow. This extra push gave the arrow a little more momentum and speed. This type of bow is called a recurved bow.

The bow's composition was, of course, also critical. The type of wood, or other material, it was made of had a large effect on its power. Also, a bow's density, elasticity, and tensile strength (amount of stress it can take before it breaks) determines how much energy it can store and how well it returns to its original shape after the shot.

Early on it was discovered that bows made from more than one material were more effective than simple wood bows. They are referred to as composite bows. Composite bows were usually made of wood, a section from the horn of an animal, and sinew. A thin section of horn was glued to the belly of the bow on the side facing the archer. Horns from antelope, water buffalo, and sometimes sheep or goats were used. This allowed more energy to be stored in the bow. The glue was made from fish oil. Strips of sinew were also glued along the back of the bow, again to increase the energy storage. The tips (recurved sections) were also stiffened using sections of bone.[8]

Arrows were continually improved over the years. One of the most critical concerns was the weight of the arrow. If an arrow was too light it would be affected by the movement of air and would not stay on course well. If it was too heavy, on the other hand, it would create a lot of drag and fall too fast. The ideal weight was somewhere in between. It was also discovered early on that feathers along the sides increased an arrow's stability, and that the length and height of the feathers had an effect on how far the arrow went, and on its stability in flight.

A variation on the ordinary bow and arrow is a crossbow, which is known to have been used by the early Greeks. It fired a steel bolt, and initially the drawstring had to be drawn back by the archer and locked into position, then released using a trigger. This made loading slow, and considerable strength was needed to pull the string back. Eventually, however, a mechanical winch system was developed for loading, and with it, a much greater tension could be put on the string. As a result, the steel bolt left the bow with a much greater velocity, and it therefore had a greater range. The crossbow could, in fact, project a bolt up to five hundred yards. But the problems didn't go away. The steel bolts were not very aerodynamic, and as a result they were also not very accurate. In addition, they were much slower and much more difficult to load than an ordinary bow. At maximum, a crossbow could be fired about twice every minute whereas a good archer could fire twelve to fifteen arrows a minute. Initially, however, crossbows had a serious advantage over the ordinary bow: the steel bolts they fired could penetrate the steel armor shielding of the enemy. Furthermore, the bolts could easily kill a horse. Eventually, however, the English invented the longbow, which also packed enough power to penetrate armor.

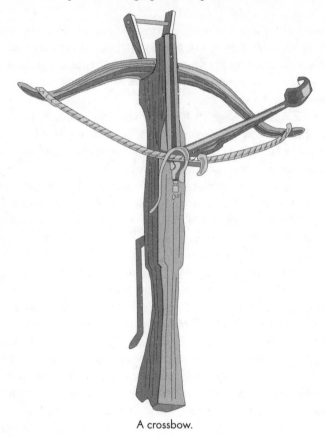

A crossbow.

Although it was not used in early wars, a significant advance was eventually made in bows. A bow is, of course, hard to draw back, but the greater the energy you expend in drawing it back, the greater the energy it transfers to the arrow. And again machines came to the rescue. Pulleys were eventually used to help archers do more work on the bow (and produce more potential energy) with less physical effort. The compound bow allows an archer to hold and aim a drawn bow without a lot of stress or fatigue.

PHYSICS OF OTHER EARLY WEAPONS

Most of the early weapons were devices of one sort or another that projected either arrows or stones. Most were, in fact, catapults of some sort. As we've already discussed, three of the best known were the ballista, the trebuchet and the onager. The ballista was a torsion spring, which stored energy in several loops of twisted skeins. It could project heavy iron-tipped darts or stone projectiles of various sizes. The darts were placed in a shallow wooden trough or groove. The ballista was armed by hooking a bowstring behind the dart and winding it back using a windlass. It pulled a sliding trough and the dart within it back, and at the same time it twisted the skein (the bow string was attached to two arms, each with its own twisted skein). Ratchets and cogs prevented it from shooting while it was being loaded. Once loaded it could be fired using a trigger. Again, we have potential energy of torsion in the twisted skeins, and when it was released, it was converted to the kinetic energy of the dart.

The earliest ballista were developed about 400 BCE. The best ones had a range of about five hundred yards. They used only relatively light projectiles, so they didn't hit with a lot of force, but they were relatively accurate.

The trebuchet was much more powerful, and it worked on a completely different principle. It is, in fact, sometimes called the "counterweight trebuchet" because it used a counterweight to produce the energy for throwing the projectile. It is said to have first been used by the French in the twelfth century, and is based on the principle of the fulcrum described earlier. Its major part is a long arm anchored above the ground close to one end. At the end of the long arm is a sling that contains a pouch; this is where the projectile (usually a large stone) was placed. The energy came from a heavy weight placed at the end of the shorter arm. It was raised up and held in place until ready. Again, we have potential energy being converted into the kinetic energy of a projectile. When the trigger was released, the sling and longer arm would swing upward toward a vertical position. At this point the sling would release and the pouch would open. The projectile would be thrown forward with considerable velocity. The device was, in effect, powered by gravity.

The advantage of the trebuchet was that it could project stones of up to three hundred pounds, which could do considerable damage to the upper parts of most castle walls. It had a range of about three hundred yards. It's important to note that the sling played an important role; it could double the power of the trebuchet, allowing it to throw a projectile twice as far as it would have been able to without the sling.

The third of these catapults, called the onager, was similar to the ballista in that it also used torsional pressure generated by twisted-sinew ropes. It consisted of a large frame on the ground with a vertical frame attached to it. A spoke, or arm, was attached to an axle that ran through the vertical frame. A bowl-shaped bucket was placed at the end of this arm, with the projectile placed in this bucket. The machine was loaded by pushing the arm back against the tension created by the twisted skeins. When the trigger was released, the arm would swing upward in an arc, releasing a projectile at the top of the swing. As in the case of the other devices, large stones were usually used as projectiles. Onagers usually had a range of about twelve hundred feet.

4

THE RISE AND FALL OF THE ROMAN EMPIRE AND THE EARLY ENGLISH-FRENCH WARS

The Roman Empire was one of the largest military powers the world had ever seen, and it dominated the Western world for over a thousand years. The Roman march to world dominance began with the Punic Wars, the first of which started into 264 BCE. These wars were fought between Rome and the other significant power in the Mediterranean, the Carthaginians. The first of the wars was fought largely at sea, and after a long struggle, the Romans persevered. In the Second Punic War (218 BCE to 201 BCE), however, the Romans came up against Hannibal, the great Carthaginian general, and he overcame them in battle after battle. Hannibal was successful mainly because he outsmarted the Romans. Confident in their ability to win, the Romans were easily led into traps, and over the next seventeen years over a hundred thousand Roman soldiers were killed as a result, but finally the Roman generals began to learn, and eventually they overcame Hannibal.

THE ROMANS AND THEIR WEAPONS

After defeating Carthage, the Romans continued their campaign, eventually conquering most of the area around the Mediterranean, Greece, much of the Middle East, Germany, North Africa, and England. Indeed, they conquered a large portion of the civilized world at the time, and in the process became the greatest military power the world had ever known. And they did it using the best weapons and tactics known at the time. They used body armor consisting of many plates of material layered on others. Most of their armor and weapons were made of either bronze or iron. They did not discover steel. They used a relatively short, but effective sword called a gladius, which was used primarily for thrusting. In addition, the Romans used bows and arrows, javelins, spears, and a shield called a scutum that was about forty inches tall, thirty inches wide, and slightly curved.

In sieges they used the ballista, the onager, and various other catapults that incorporated springs of various types. Their standard tactic when facing an enemy was to advance with their shields side by side so that they were less vulnerable to enemy fire. The front line, however, was rotated every fifteen minutes to keep it fresh. The soldiers underwent harsh and grueling training, and discipline was strictly enforced.

Although they used the ballista, the onager, and other weapons that were used by the Greeks, they did not try to improve on them. In fact, the Romans made almost no military advances in weaponry. When they noticed that an enemy had something innovative and new, however, they quickly adopted it. Indeed, they appeared to have no interest in science, either in its own right or as something that might be used to develop new weapons. For the most part, they had a disdain for science. When they conquered Greece and Alexandria, hundreds of thousands of scientific texts came into their hands, but they never went to the trouble of translating any of them, and they never made use of anything that was in them. Their view was that they had all they needed to win wars.[1]

Surprisingly, though, despite their disdain for science, they did excel in engineering. Over the hundreds of years they were in power, they built thousands of miles of roads—some of the best ever seen in Europe. Furthermore, they built a large number of dams along with networks of aqueducts to distribute the water throughout the land. And they built some of the largest and most magnificent buildings seen up to that time. Many of them were based on the arch. They were also excellent bridge builders. So, while they knew little about physics and cared less about it, they actually used many of the basic principles of physics. For their engineering feats they had to understand such concepts as force, weight, stress and strain, and water pressure.[2]

The Roman Empire lasted from about 753 BCE to about 486 CE. It was at its peak in about 250 BCE, when it had conquered most of the known civilized world. But its large size eventually caused things to unravel. It was difficult for Rome to oversee so many distant lands and people. In addition, there was strife at home: there were power struggles among the generals, and as a result, several civil wars eventually broke out. After Caesar was assassinated, Octavius and Mark Antony fought a civil war against the assassins, but eventually the two men began to fight each other. Furthermore, because of the extent of their large empire, they began incorporating troops from their conquered lands into their army, along with mercenary troops, until finally these troops came to represent a large fraction of the Roman force. This created a significant change in the army;

no longer did soldiers undergo rigorous training and discipline, and their dedication to Rome decreased significantly.

The only enemies they had, however, were at the boundaries of their empire, and these enemies were barbarians who knew nothing about organized warfare, siege engines, and body armor. And to Rome they were not a significant threat.

But while the Roman army was declining, the barbarians were looking for something that would give them an advantage. And as we saw earlier, the answer was the use of warriors on horses. Most barbarians learned to ride early. The biggest problem was staying on the horse during a fight—in particular, controlling the horse while shooting arrows and using swords along with lances and spears. The first breakthrough was the development of the saddle; the first simple saddle was devised by the Scythians of Eastern Europe. It was a crude affair with horsehair used as cushions in front of and behind the rider. Cloth loops were soon added to it for the rider's feet. Later, a harness was added to give the rider much better control of his horse. Finally, the cloth loops were replaced with iron stirrups, which were particularly helpful in stabilizing a warrior during battle. There is, however, some controversy as to whether the stirrup was used before 376 CE.

The battle of Adrianople, as it was referred to, had its beginnings in 376 CE. One of the barbarian tribes, the Goths of modern-day Turkey, sent a message to the Roman Empire asking if they could occupy some of the conquered Roman land near the Danube River. Thinking they would be good soldiers, the Roman generals allowed them to occupy it. But it turned out to be a mistake. Fighting soon broke out between the Goths and the Romans in the area. Over the next few years, in fact, there were several battles.[3]

The Roman emperor, Valens, became annoyed and wanted to end the problem once and for all. In 378 he decided to go to Adrianople himself. Leading an army of about thirty thousand troops, Valens arrived in the area of Adrianople in early August. He was informed that an army of about ten thousand Goths, led by the chieftain Fritigern, was marching toward Adrianople and was about twenty miles away. Valens continued on to Adrianople and set up a fortified camp, ready for Fritigern and his army. He was informed that more reinforcements were coming to help him, but he didn't want to wait for them despite strong urging from his generals. He was sure the strong Roman army would easily overcome the barbarians.

Fritigern sent a message to Valens proposing a peace and an alliance in exchange for some Roman territory. Valens was so sure of victory that he rejected the proposal and prepared to charge. As it turned out, Fritigern was just stalling for time; he was waiting for five thousand more highly trained cavalrymen. To further delay the attack, Fritigern sent troops to set fire to the fields between the armies, and he also tried to initiate negotiations for the exchange of hostages. This did delay the attack, but it also annoyed Valens.

Suddenly a detachment of overanxious Romans began the attack without orders, but they were easy easily pushed back by Fritigern's men. By then, however, it was too late to stop things; the other contingents of Romans continued the attack and broke through the Goth's circle of wagons. At this moment, however, the five thousand highly trained cavalrymen arrived from all directions and encircled the Romans. The cavalrymen were mounted on large, strong horses, and they had developed an extremely effective lance. The weight of the horse behind the lance made for a lethal combination, and the Roman shields were of little help. The Romans also had cavalry, but it was no match for the Goth cavalry. Many Roman cavalrymen deserted.

The Roman troops were soon in disarray. Over the next few hours the Goths slaughtered them. It was one of the greatest defeats the Roman army had ever experienced, and it was a devastating blow to the Roman Empire. The core of the army in the eastern section of the empire was destroyed. The biggest shock, however, was that the Romans were no longer invincible.

Furthermore, many of the important generals of the Roman army were killed. Valens's fate is not known, but according to one story he escaped the battlefield with some bodyguards and hid in a peasant's cottage. Fritigern's men attacked the cottage, and Valens's men tried to defend it by firing arrows at them. But the Goths set fire to the cottage and Valens perished in the flames.

This battle is generally considered the start of the final collapse of Rome, and there's no doubt that it had a serious effect. It showed that the Romans were not invincible. Nevertheless, the empire lasted for another hundred years. During this time, however, it continued to shrink in size. Most of the attacks during the late years of the empire were mounted by another barbarian tribe, the Huns. But they used the same tactics as the Goths.

EARLY ENGLISH-FRENCH WARS

After the fall of Rome in 476, the world entered a period now referred to as the Dark Ages. Few advances in science were made during this time. In addition, relatively few historical and other written records were left during this period, compared to earlier and later periods. The Dark Ages lasted from 476 to about 1500 CE. During this time barbarian tribes—the Mongols, Huns, Goths, and even Vikings from the north—swept across Europe. Physics, and science in general, was virtually at a standstill for years, but one science did flourish, namely metallurgy. Mounted warriors were now the norm, and they were soon armored. Cavalrymen eventually had a full suit of armor made from small metal rings overlapping one another, referred to as chain. Later, because archers frequently shot the horses out from under them, horses were also armored. As a result, there was a continuing search for metals that would give better protection. And soon after cavalrymen found that arrows were penetrating their chain armor, steel plating was developed.

For many years the armored knight on a horse with a long lance or sword was in the lead when a charge was made. Not only were such mounted knights very effective, but they also had considerable shock value. Few infantrymen would stand their ground against a knight on a horse as it barreled down on them. Furthermore, the speed of the horse was much greater than the speed of a man on foot, so the infantrymen had no place to go. This, along with the tremendous thunder of the horses' hoofs, no doubt had a huge psychological effect.

The main problem with the knight was the expense. Their armor came at a high price, and when "chain mail," as it was called, was found to be vulnerable to arrows, metallurgists had to go to work. Steel was soon developed, however, and small steel plates were placed over the chain mail. But even the steel plate eventually became vulnerable to the crossbow, and later to the longbow.

The crossbow fired steel bolts. The biggest problem with early crossbows was that it took a lot of strength to load them, and, as a result, they had a slow rate of fire. Later on, however, in the eleventh century, a mechanical winch system was added that overcame the problem of loading the weapon; furthermore, it allowed archers to create much more tension in the string, giving the bolt considerably more kinetic energy and velocity when it was released.

The drawback of the crossbow was that steel bolts were not as aerodynamic as arrows, and they were therefore less accurate. Nevertheless, the crossbow became a real threat to a knight armored only with chain mail.

The tactics of this period are seen in one of the most important battles to occur at the time: the Battle of Hastings, which took place in 1066. It was fought between an English army under King Harold II and a Norman army under Duke William II, and it took place about six miles from Hastings in England.[4]

The tactics of William proved to be much more effective than those of Harold. He had a force of archers, cavalrymen, and infantry working together; Harold's army was almost entirely infantry. Both men had about the same number of fighters—approximately twenty thousand. The English army, mostly infantry, used a conical metal helmet, a metal vest, and a shield; its primary weapon was the ax, but some of the men used swords.

William's army attacked first, shooting volleys of arrows, but most of them merely hit the shields of the English and had little effect. Believing this had weakened the English, William charged, but the English threw everything they had at his forces, and the English inflicted many casualties in hand-to-hand fighting.

Suffering heavy casualties, the Normans retreated. The English broke rank and followed them, and the fighting became fierce and confused. William's horse was killed and he toppled to the ground, but he was not killed. He somehow managed to rally his troops and counterattack, but he now realized that the initial hail of arrows had had little effect, so he ordered his archers to fire over the men with shields in the front line to target the unprotected rear ranks of the English army. This had a considerable effect. Indeed, Harold himself was struck in the eye by an arrow.

The English were beginning to weaken and break ranks behind the wall of shields. So William ordered another attack, and this time the Normans broke through and finished off Harold. Without a leader, many of the English began to desert and flee. Soon the fight was over and the English had a new king; William was crowned on Christmas day, 1066, at Westminster Abbey.

The main thing the battle demonstrated was the effectiveness of archers — the Norman army had taken the field with about eight thousand.

Disagreements between the English and the French continued for years. One of the major problems was that William, now the king of England, also remained duke of Normandy. Thus, English kings owed homage to the king of France. But in 1337 Edward III of England refused to pay homage to Philip VI of France, and this led to a series of wars that lasted for over one hundred years (from 1337 to 1453). Known as the Hundred Years' War, this period yielded one of the greatest advances in weaponry. It first appeared in the Battle of Crécy in 1346, and it was called the longbow.

The English army was led by Edward III, and the French army was led by Philip VI. The French army was about twice the size of the English army, which consisted of about 5,000 foot archers, 3,250 mounted archers, and 3,500 other troops. The French had about 6,000 crossbow men as well as a large infantry. The battle took place near the forest of Crécy.[5]

Philip placed his crossbow men in the front line, with the cavalry behind. Strangely, they left their wooden shields (their only defense) in carts at the rear of the formation. The battle began with a series of volleys from the French crossbow men, which proved to inflict little damage. The English had a new longbow, and while the crossbow men could only shoot at the rate of one to two shots per minute, the longbow men could fire five or six shots in the same time.

Furthermore, the longbows had a much greater range and a deadly penetrating power. To make things worse, the crossbow's strings had been weakened by a rainstorm just before the battle. The longbow men had unstrung their bows and protected them.

The hail of crossbow darts fell short, but the barrage of English arrows that followed didn't, and it was effective. The barrage continued, and, as a result, the crossbow men couldn't get close enough for their bows to be effective. Seeing their situation, many of the crossbow men began retreating through the lines of French knights behind them. Angered by the retreat, the knights began hacking at their own men, killing many of them. Then they decided it was time to charge. They forced their horses forward over the retreating crossbow men, trampling many of them. The English bowmen continued firing at the knights, and to the surprise of the French, many of the arrows penetrated their armor. As a result, many of them began to fall, and as more and more fell, they began blocking the men behind, and bodies began falling upon bodies.

When the battle was over the French had suffered tremendous casualties. According to one estimate, four thousand French knights were killed and as many as two thousand French archers. The English, on the other hand, suffered few casualties—most estimates were under three hundred. The decisive factor was the new longbow, and it would continue to play an important role for another hundred years or more.

A longbow of the type used at Agincourt.

The longbow was also decisive in the Battle of Agincourt in October 1415. By this time it had become even more effective. Again the English and the French were involved; this time the English army of about six thousand men was led by Henry V, and the French army numbered twenty-five thousand or more. The battle was fought on a narrow strip of open land near Agincourt.[6]

With a four or five to one advantage, the French were likely overconfident. They had eight thousand heavily armed men, but they would only be effective in hand-to-hand battle, so they had to get close. Furthermore, between the two armies was a recently plowed field, and it had rained heavily during the previous days. So it was muddy, much to the detriment of the French, many of whom were heavily armored. But on the English side, many of the troops were sick and exhausted from days of marching.

The English archers drove long, sharp stakes into the ground at an angle pointed toward the French line. This helped protect them from cavalry charges by the French knights. The French formed up in three lines, with the men-at-arms in the front and the archers and the crossbowman behind. The French were expecting the English to launch a frontal attack, but they didn't. Instead, the English longbow men opened up with a barrage of arrows. The well-trained longbow men could now fire up to fifteen arrows a minute, so within seconds there were thousands of arrows in the air. Furthermore, to the surprise of the French, the arrows easily penetrated their armor. Many of the arrows struck the horses on their backs and flanks, causing them to panic; as a result, wounded and panicked horses galloped through the advancing infantry, trampling them.[7]

The French men-at-arms tried to protect themselves as they moved forward. Because their helmets were the weakest part of their armor, most of them lowered their heads to avoid getting shot in the eye or through the breathing holes in the helmet. This restricted their vision. Along with this, they had to walk through knee-deep mud in places, and also over and around fallen comrades. And the barrage of arrows seemed to be unending. Soon the French ground troops were exhausted and disheartened; furthermore, they couldn't get close enough for hand-to-hand fighting.

To make things worse, the second and third line of French warriors didn't know what was happening up front, and they continued to forge forward, and soon they suffered the same fate. The battle lasted three hours, and in the end four thousand to ten thousand French were dead (according to various estimates), with English casualties as low as a few hundred. Many of the elite, including dukes, constables, royals, and so on, were killed. And again, it was the English longbow that was the decisive factor.

ORIGIN AND PHYSICS OF THE LONGBOW

The longbow was developed in several countries independently. In Great Britain it was first developed by the Welsh. And there's no doubt that they made significant advances in its construction, not from understanding any of the science behind it, but mostly from trial and error.[8]

The English felt the effects of the Welsh longbow early on. It was used against them, mostly in ambushes and skirmishes at first, but eventually in larger battles, such as one in 1402 where the Welsh used it quite effectively against the English. This, of course, caused considerable concern for the English, and it also piqued their curiosity. They soon incorporated Welsh archers into their own army and learned their techniques.

The first English longbows were made of a single piece of wood—usually yew because it was particularly springy and sturdy. The major problem was that yew was not a common tree and was relatively rare in England. Because of this, they were sometimes made of elm or ash.

Staves were selected with great care and went through a relatively long production process. Oil and wax were applied to the stave to make it waterproof and help preserve it. It had to be relatively thin, and the length would be customized to the archer. The longest measured about six feet four inches, with shorter ones somewhat over five feet. Since there was a direct correlation between the bow length and the power it could generate, the longer the bow, the better. It was usually about two inches across at its thickest part. The force required to pull a longbow back to its maximum extension ranged from about eighty to one hundred twenty pounds. The draw length was from twenty-nine to thirty-two inches, and it was soon determined that it worked best when the bow was drawn back to the eye.

Arrows were made of a variety of woods. Aspen, poplar, elder, willow, and birch were all used, with the average length of an arrow being about three feet. It was determined early on that feathers along the side helped to stabilize the arrows in flight, and the feathers were usually seven to nine inches long, and glued to the shaft. The bowstring was usually made from hemp, but later on flax and silk were used.

One of the major problems with the longbow was the training required to master it. Because of the tremendous force required to pull it back, considerable practice was required to use it effectively, particularly in battle. As a result, English boys usually began their training by about age seven. They were trained extensively, and tournaments were held in all villages, with the best archers being selected for the military. And it was a great privilege to serve as a military archer, as the archers were considered to be members of an elite group.

The average trained English archer could fire at least twelve arrows a minute and hit targets at two hundred yards. Indeed, if you could fire at a rate of only ten arrows a minute you were considered a poor archer.

We looked at the physics of the bow and arrow briefly earlier, and much of what we said also applies to the longbow. We'll consider it in more detail, however. The physics involves both the mechanics of the bow and the flight of the arrow. As we saw, when the archer pulls back the bow he does work that is stored as potential energy. When the string is released this potential energy is converted to the kinetic energy of the arrow. Actually, some of the potential energy goes into the final motion of the bow (a slight vibration), but it is usually a small portion. It is important to note that the farther the string is pulled back for a given bow, the greater the potential energy. This is why the longbow was able to impart more kinetic energy to the arrow. It is longer and can therefore be drawn back farther.[9]

The range, or distance, the arrow travels, depends on the following things:

- Initial velocity
- Weight of the arrow
- Angle at which the arrow is shot
- Air resistance
- Effect of the wind

The arrow's initial velocity can be determined by equating the potential energy (F × d) of the bow (with the string pulled back) to the kinetic energy of the arrow ($1/2\ mv^2$, where m is the mass of the arrow). The angle at which the bow is pointed has a strong bearing on its flight path, or trajectory, and how far it will go. It's relatively easy to show that the greatest range is obtained an angle of forty-five degrees if air resistance and wind are not taken into consideration. But as we will see later, these variables are also important, and they do limit the range.

The path of the arrow is a parabola. This is the curve seen in the headlight of a car. But because of air resistance, it can be a slightly distorted parabola. Air resistance creates a force on the arrow that slows it down; this is the result of a transfer of some of the momentum of the arrow to the air. There are two types of drag on the arrow: sheer drag and form drag. Sheer drag occurs because the arrow drags the air adjacent to it along with it as it moves. Indeed, if you could closely examine the arrow in flight, you would see that there is a series of layers of air around it, with the layer closest to the arrow being dragged the most, the second layer being dragged to a lesser degree, and so on. Sheer drag is proportional to the velocity of the air moving past the arrow.[10]

Form drag occurs because sheer drag causes eddy currents behind it. These eddy currents form a turbulent wash behind the arrow in the same way that a speedboat creates a wash when it moves through water at high speed. And the faster the arrow goes, the greater the turbulence and the greater the form drag. Mathematically, form drag is proportional to the square of the velocity (v^2). It acts in a perpendicular direction, pushing the arrow to the side and creating frequency oscillations during its flight.

Furthermore, when an arrow is released, it creates a perpendicular kinetic energy. For a right-handed archer, when the arrow is released the string moves slightly to the left, causing the arrow to bend to the right. Then the string moves back to the right, causing the arrow to move left. All this happens in the brief time that the arrow is still attached to the string. But when the arrow leaves the bow, this slight right-left vibration continues through the flight. If the archer is left-handed, the vibration is opposite.

The amount of vibration depends on the stiffness of the arrow. If it is quite flexible it will vibrate excessively, which in turn will cut down on its speed and therefore its penetrating power. If it is too stiff, on the other hand, the arrow will not vibrate, and this affects its accuracy. So compromise is therefore needed.

The effective range of the longbow in the hands of early archers was generally about two hundred yards, and at that distance it could easily penetrate the chain mail armor used by most knights. Eventually, steel plates were placed over the chain for greater protection. But the longbow arrows could penetrate even the steel plates if the range was less than one hundred yards. The maximum range of the longbow was about four hundred yards. At the beginning of a battle, archers would usually shoot large numbers of arrows high in the air so that charging knights would encounter thousands of arrows falling from the sky. Later, as the charging knights got closer, the archers would select individual targets.

5

GUNPOWDER AND CANNONS

The Discoveries That Changed
the Art of War and the World

For years Genghis Khan, the leader of the Mongols, had been eyeing China. It was prosperous and thriving, and the Chinese had many things that the Mongols wanted and needed. In 1205, Genghis Khan decided to attack. It was little more than a raid, but it frightened and panicked the Chinese. The Mongols were cruel and well known for their use of terror; they took no prisoners and frequently wiped out entire villages.[1]

The Chinese had to do something, and they had to do it fast. They needed a weapon to counter the Mongols, and indeed they had something with considerable potential: a white powder that they were using in their fireworks.

Genghis Khan attacked again in 1207, causing even more fear. This attack spurred the Chinese to develop what would become known as the "fire lance." It was a bamboo tube several feet long that had been drilled through the joints, with the bottom joint left in. The tube was wrapped to reinforce it and a small "touchhole" that could be used as a fuse was drilled near the lower end. White gunpowder was poured into the bottom, and arrows or other projectiles were placed on top of it. When the gunpowder was lit through the touchhole, the arrow left the barrel with considerable speed, but it only had a range of about ten feet. They also developed other weapons, such as simple flamethrowers, rockets of various types, bombs that could be thrown with a catapult, and land mines.

The Mongols declared full war in 1211 and swept down on horseback by the thousands, but the Chinese fought gallantly with everything they had. For two years the Chinese held them off, but eventually the Mongols overcame them, and soon they were using the new Chinese weapons against others.

Over the next few decades the art of war changed significantly. A revolution had begun, but at this time no one realized how important gunpowder would eventually become in warfare. To tell the entire story, however, we have to begin with the discovery of the components of gunpowder, one of the most important of which is saltpeter. Saltpeter is actually potassium nitrate, and at the time of the Mongol invasion it was relatively rare in China. One of the first places it was found was on the walls of caves, where it could be scraped off as a white crystalline powder. It generated interest because it flamed up when sprinkled on a fire. Later it was also found on the floor of stables that housed horses. Alchemists showed that it came from the urine of horses.

Genghis Khan.

The biggest problem with the early form of saltpeter was that it was actually a mixture of potassium nitrate and calcium nitrate. Techniques for improving its purity were developed over the years by alchemists, and they eventually became interested in it as a possible elixir of immortality. In the early 800s, however, they began mixing it with other chemicals. A mixture that soon began to attract interest was the mixture of saltpeter, sulfur, and carbon (in the form of charcoal). When rolled in paper and set on fire it caused a loud explosion, and as a result it was soon used in celebrations and to scare away evil spirits.

The Chinese alchemists no doubt started with a 1:1:1 mixture of the three materials, but they eventually found that a 4:1:1 mix (with the 4 corresponding to saltpeter) gave a much better explosion. As it turned out, saltpeter was the oxidizer in the combination (so it didn't need air to explode), with sulfur and carbon acting as fuels. The ratio between the three components was, indeed, critical, and it was later found that different ratios yielded even greater explosions. For gunpowder, however, the three major components remained the same.

For a few hundred years after the discovery of gunpowder there is no indication that the Chinese used it for anything but celebrations and as "toys" for children. But all that changed when the Mongols attacked. At first they over-came a few villages in the north, but finally they not only conquered all of China, but also continued on their rampage throughout Europe, eventually cap-turing most of it. And not only did they use the new Chinese weapons, but they developed others and used them in their further conquests.[2]

News of the new weapons spread rapidly. By 1250 the Arabs had begun to use gunpowder in a simple "cannon" they called the madfaa. It consisted of a wooden bowl or pot that was packed with gunpowder, with arrows or other projectiles such as stones above it. When the fuse was lit the arrows or stones would fly in the general direction of the enemy. Needless to say, the madfaa was very inaccurate.[3]

ROGER BACON

The news of the strange new explosive eventually reached England in the mid-1200s, and an English philosopher and Franciscan friar named Roger Bacon heard about it. A trader or missionary brought him a "firecracker" that had been made in China. Bacon had a strong interest in science; indeed, he eventually made significant contributions to many branches of science and mathematics, including optics and astronomy. Furthermore, he had a relatively good knowledge of chemistry. He took the firecracker carefully apart and examined the powder inside. Analyzing it, he found it to be made up of saltpeter, sulfur, and charcoal, and it didn't take him long to realize it was an important mixture that could have significant application in war. He worried that it might get into the wrong hands; nevertheless, he mentioned the discovery in his book *Epistolae de Secretis Operibus Artis et Nature et de Nullitate Magiae* a few years later, though he was reluctant to publish the exact formula for it. It is said that he finally decided to publish it in the form of a cryptogram, but others have disputed this.[4]

One of the problems with gunpowder at this stage was that it depended criti-cally on saltpeter, and Bacon soon discovered that the saltpeter in the mixture was impure. He therefore made a study of it and found a way to purify it.

While Bacon was making his discoveries, the Chinese were still working on their weapons. Within a few years they had developed a crude form of a cannon, but at this stage it was difficult to use and dangerous when fired. And, of course, their crude cannons were terribly inaccurate. They noticed, however, that if the explosive gases were contained in a fireproof and blast-proof chamber, the resulting "explosive power" was significantly increased.

Evidence that the Chinese were working on gunpowder came in 1280 when a large gunpowder arsenal caught fire. The resulting explosion was heard for miles around, and it was reported that over one hundred guards were killed in the explosion. Furthermore, pieces of the storage building were found over two miles from the explosion.

DEVELOPMENT OF THE CANNON

Crude early cannons were made by the Chinese, the Arabs, and the Mongols, but the first cannons as we know them appear to have been built in Germany and Italy. The name *cannon*, incidentally, comes from the cylindrical barrel of the device; the Latin word for it is *canna*. The Latin word *canon* was also used later to mean "gun."[5]

An early Chinese hand cannon.

The first English cannon appeared in 1327, and there is some indication that the German Berthold Schwartz (sometimes called "Black Bart") mixed up the ingredients of gunpowder about this time and may have made a simple cannon. According to folklore, he put his mixture in a pot and covered it with a large stone slab, and somehow a spark ignited it and blew the large slab through the roof of his laboratory.[6]

In 1326 the first drawing of a cannon appeared in a book by Walter de Milemete. And in 1341 a poem titled "The Iron Cannon Affair" was published by Xian Zhang. He wrote that a ball fired from a cannon could "pierce the heart or belly . . . and can even transfix several persons at once."[7]

Cannons were still relatively rare in Europe in the 1340s, however, and they still shot only arrows and grapeshot. In some cases the arrows were aflame in flight. Furthermore, in the earliest cannons the gunpowder was a fine powder and the resulting explosion was not impressive. Indeed, only a fraction of the gunpowder actually exploded. The gunpowder was placed in a cylinder of bronze or iron, which was closed at the lower end, with the cannonball placed on top of it. Large stones were initially used as cannonballs, but iron balls eventually replaced them. They were spherical because of their tumbling motion in flight. The gunpowder was ignited through a touchhole. In a sense it was a fuse, but it actually consisted of a narrow stream of gunpowder. When the gunpowder mixture was ignited it quickly changed from a solid to a gas, which increased its volume by a factor of about four thousand. The gas was highly compressed and created a tremendous force as it expanded, pushing the cannonball down the barrel.

A lot of science and technology is required in the building of a cannon. Metallurgy is important in the building of the cannon barrel and housing; chemistry is important in determining the best mixture of materials for the gunpowder and how much to put in the cannon. Too little and the ball will not go far; too much and the cannon will explode, killing anyone near it. In these early years, in fact, many soldiers were killed by exploding cannons.

The drill that a cannon crew went through each time a shell was fired was somewhat as follows:

a) The correct amount of gunpowder had to be poured in the barrel, and it had to be tamped down.
b) A wad or plug was usually placed over it.
c) A cannonball was placed in next and pushed firmly against the pad.
d) Gunpowder was poured in the touchhole.
e) A flame was brought up to the touchhole.

THE HUNDRED YEARS' WAR

The cannon was first used by the English in the Hundred Years' War, which started in 1337 and ended in 1453. It was a long war that was fought off and on between the French and English. Most people remember it for the important victory of the French over the British when the French were led by Joan of Arc, a seventeen-year-old peasant girl who was burned at the stake when she was only nineteen.

Cannons were used very little in the early years of the war. At this stage they were still inferior to siege engines; they couldn't penetrate castle walls, and they had many problems. By the time of the last battle at Castile in Gascony in 1453, however, cannons had been improved significantly, and they proved to be quite effective. Three hundred cannons were used in this battle.

THE BASILICA AND THE SIEGE OF CONSTANTINOPLE

Cannons also played a critical role in one of the most important battles of this period—the battle that led to the downfall of Constantinople in 1453. For years there had been tension between the Ottoman Turks and the Byzantine Empire. Then, in 1451, Sultan Mehmed II came to power. Many thought he was too young to be taken seriously, but he soon proved them wrong. Within two years he had a large army and was preparing to attack the Christian stronghold at Constantinople, a city that was protected by a huge fifteen-mile-long wall that was thought to be impregnable.[8]

But Mehmed was determined. Under his direction, one of the largest cannons ever built, called *Basilica*, was constructed. It was designed and built by a Hungarian engineer by the name of Urban. Little is known about him, but there is no doubt that he knew a lot about cannons. Initially, Urban offered his services to Constantine XI of Constantinople, but he was turned down (records indicate that his asking price was too high), so he went to Mehmed and the Ottomans and they accepted his offer.

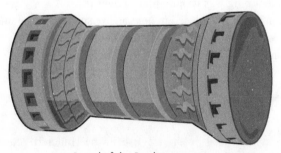

Barrel of the Basilica cannon.

Basilica had a barrel twenty-seven feet long, which was longer than any cannon at that time, and it was said to be able to a project a six-hundred-pound stone over a mile. It was so large, in fact, that it took three hours to load it after each firing. But it was one of the first cannons that had enough firepower to break through a castle wall. And it wasn't the only cannon the Ottomans had. When their army lined up outside Constantinople they had sixty-eight cannons pointed at the walls. Most were considerably smaller than Basilica, but because of their number they could inflict a lot of damage.

For fifty days the Turks bombarded Constantinople with their cannons. And indeed, under the tremendous bombardment the walls began to crumble in places. They were having an effect, but the soldiers within the walls continued to repair them as quickly as they were penetrated.

Inside the city there were seven thousand defenders, while outside Mehmed had one hundred thousand to one hundred fifty thousand soldiers ready to attack once the walls were down, so the odds were strongly in his favor. But Mehmed and his men still had to get through the wall. After fifty days of bombardment Mehmed began to get impatient. Considerable damage had been done, so he sent a contingent of soldiers to see if they could break through the weakest section of the wall. The Christians fought them off, however, and many of his men were killed.

Still determined, Mehmed sent another contingent of men in a much more organized attack. They managed to actually enter the city, but again the Christians were ready for them, and, after a tremendous fight, Mehmed's surviving attackers were forced to retreat.

Finally, at about midnight on May 29, Mehmed ordered an all-out attack. Among his troops were his favorites, the highly trained Janissaries. They used everything they had as they attacked, but it seemed again that they might be overcome. Then their luck turned: someone noticed that one of the gates had been left unlocked. Some of the soldiers managed to get in and unlock some of the other gates. Mehmed's men were soon flooding into the city. Mehmed also had several ships sitting in the harbor with many more men. When they heard the news, they, too, joined the fight, and soon it was over. After fifty long, hard days, Constantinople had fallen. Mehmed claimed the city for Islam and renamed it Istanbul. Over the next few years he built numerous temples, mosques, and shrines within it.

CANNONS IN THE ENGLISH-SCOTTISH WARS

The English used their first cannon in 1327 against the Scots, and they also used it in the Hundred Years' war. The Scots first used a cannon in 1341 to defend a castle. James II of Scotland was particularly enthusiastic about cannons. He used them to attack Roxburgh Castle in 1460; it was the last Scottish castle that was held by the English, and he was determined to get it back.

In the attack he used a large cannon nicknamed "the Lion." While he was standing beside it, however, it exploded and killed him. Nevertheless, the Scots continued the siege, and within a few days they had overcome the English and reclaimed the castle.

One of the most famous Scottish cannons was called *Mons Meg*. It was sent as a gift to James II in 1457, and is still on display at the Edinburgh castle today. It had a length of over thirteen feet and an inner diameter of twenty inches, and it fired cannonballs that weighed four hundred pounds. For some time it was the largest cannon in the world.

THE FRENCH

After their defeat to the English in 1415 at Agincourt, the French knew they had to find a defense against the English longbow. Soon after he came to power, Charles VII decided to do something about it. He vowed to expel the English from France, but he knew that a new approach would be needed. He therefore recruited the best minds in the country—particularly engineers and physicists—and put them to work. The cannon obviously had considerable promise, but there were many problems with it. At the time cannons were large, heavy, and difficult to move, and it appeared that the only way to improve their performance was to build bigger and bigger ones. Still, even the largest ones were not very efficient at breaking through castle walls. Charles set his team to work on the problems.

The best cannons at the time were bronze, but they were expensive. The team began experimenting with cast iron and various other types of iron. They also soon determined that they needed a longer barrel to allow for a greater controlled expansion of the exploding gases. They improved the cannonballs by making them slightly smaller than the cannon bore so that they fit rather tightly into the gun, but with enough room so that some of the gases escaped as a safety measure. They also tapered the barrel from the base to the mouth.

The problem of recoil had been around for years and no one knew what to do about it. At first artillerymen tried to tie down their cannons, but the force of the recoil always broke any shackles they used. We now know, of course, that recoil is explained by Newton's third law, but the artilleryman of the day knew nothing of Newton's laws of motion. They finally discovered that the best solution was wheels: let the cannon recoil and fly backward on wheels, then pull it forward again quickly for the next shot.

Wheels also solved another problem: maneuverability. Cannons were heavy and not easy to move. Wheels solved this problem. And while they were able to do some damage to castle walls, they were not terribly effective. Only the super cannons that were incredibly heavy could do much damage. Though ignorant of the finer detail of the physics of momentum and impact, Charles's team eventually came to realize that it wasn't just the mass of the projectile that mattered. The projectile's speed as it struck the wall was also critical. The faster, the better. So they experimented with gunpowder, trying to make it more efficient, and they found that granulated gunpowder was better.

Finally, the trajectory of the projectile was a problem. At first they believed that the cannonball flew up into the air, and that at some point it dropped straight down. Although they got some insight into this problem, it wasn't solved for many years. Also, they determined that the angle at which the cannon was aimed was of critical importance. It determined where the cannonball would land. Different angles meant different ranges, so they designed the trunnion. It used ropes, wedges, and screws of various types to allow the cannon to be pointed at different angles.

Finally, Charles VII was ready, and soon he had driven the English out.

CHARLES VIII AND VICTORY OVER NAPLES

Charles VIII took advantage of the many advances made by Charles VII. In the late 1400s he assembled a huge army and decided to attack some of the cities in Italy. In 1494 the duke of Milan and others encouraged him to attack Naples. So Charles, with an army of twenty-five thousand men set out for Naples, reaching the outskirts of the city in February 1495. Between him and Naples, however, was the huge fortress of Monte San Giovanni. Everyone was sure it was impossible to get by it. It had high walls, several feet thick, and over the previous hundred years its defenses had not been overcome despite numerous tries.[9]

Charles wheeled his new cannons to within one hundred fifty yards of the walls and began firing fifty-pound iron balls at it. The defenders were sure they were too small to do much damage. But Charles knew better. The guns continued to batter the walls for eight hours, and indeed, they fell. Within hours the battle was over, and Charles had lost no men. He then marched on Naples itself and took it without a battle. Charles then crowned himself King of Naples.[10]

News of the siege soon spread across Italy. People throughout the land were in shock. The leaders of the Italian city-states had to act fast; they needed a defense against the new weapons. Someone soon found that if they piled dirt behind the walls, the shelling was much less effective and would do little damage.

The critical factor over the years was science, particularly physics and chemistry. But in reality, aside from war, there was little interest in science for its own sake. Astrology and alchemy were still ingrained in society. Kings had professional astrologers and alchemists in their courts, but not scientists. Mysticism still ruled.

6

THREE MEN AHEAD OF THEIR TIME

Da Vinci, Tartaglia, and Galileo

The Dark Ages lasted until about 1500, but by the late 1400s important advances were being made, not only in military technology but also in the understanding of nature. One of the first in this era to tackle many of the basic enigmas of nature was Leonardo da Vinci, who lived from 1452 to 1519. He didn't have a significant impact on the science of his generation, however, mostly because few of his ingenious inventions were actually built, and his recordings and drawings were not published during his lifetime. Nevertheless, he is now considered to be one of the greatest geniuses who ever lived, and there's no doubt that many of his ideas were years or even centuries ahead of his time. Today he is best known to most people as a painter. Almost everyone is familiar with his two best-known paintings: *Mona Lisa* and *The Last Supper*. But he was also a master engineer, an inventor, and a scientist with an incredible curiosity about nature and a tremendously inventive imagination. He was, indeed, a genius of the highest caliber, and his studies spanned a wide range of areas, including physics, astronomy, mathematics, optics, hydrodynamics, chemistry, and anatomy.

He was born in Italy in the small town of Vinci, near Florence—the illegal son of a wealthy notary and a peasant girl. For the first five years of his life he lived with his mother, and then he went to live with his father, uncle, and grandfather. Although his father didn't have a strong influence on his intellectual life, both his uncle and his grandfather did. His uncle instilled a curiosity about nature and science, and his grandfather kept journals in which he wrote about his life and the events of the day. Leonardo picked up this habit from him and recorded his thoughts, inventions, and so on in journals he kept almost daily throughout his life.[1]

In 1466, at the age of fourteen, Leonardo was apprenticed to the renowned painter Verrocchio of Florence, one of the best artists in Italy. Leonardo learned a lot during the years he spent with Verrocchio. In 1478, at the age of twenty-six, however, he left Verrocchio's studio, and also his father's house, and went out on his own. He was now qualified as a "master" in the guild of artists, but he found it difficult to get work as an artist. Fortunately, he had developed another talent. Over the years he made many sketches in his journals of new types of weapons. Because military engineers were in demand, he went to Milan and applied for a job with Duke Ludovico Sforza, the ruler of Milan, but Sforza, unimpressed by his futuristic and fanciful drawings of weapons, turned him down. Nevertheless, he stayed in Milan and worked there from 1482 to 1499. During this time he continued his studies of science and engineering devices, and he rapidly became a talented inventor.

In 1494 Charles VIII of France attacked Italy. When his army reached Milan in 1499, Leonardo fled to Venice, and when Venice came under attack he soon found employment as a military engineer responsible for inventing new and better methods for defending the city.

Leonardo da Vinci.

In 1502 he was employed by Cesare Borgia, again as a military engineer. Borgia put him to work making a detailed map of the area around his stronghold. Maps were new at the time, and very few maps of any sort existed. Leonardo threw himself into the job with enthusiasm, stepping off distances and so on carefully. His map impressed Borgia so much that he immediately made Leonardo his chief military engineer.

By now Leonardo had many pupils, apprentices, and followers, and in 1506 he returned to Milan with most of them. Later he moved to the Vatican in Rome.

LEONARDO AND PHYSICS

Leonardo was meticulous in taking notes. His approach to nature and science was mainly one of observation and study, but his imagination was always at work. He spent hours studying the flow of water under various conditions. He noted how it flowed around barriers of different types, and how it varied in speed as it moved. He was particularly interested in turbulent flow and the dynamics it created. And from what he learned he developed many devices that used the force of water.

He also had a fascination with the flow of air around objects, and he invented an anemometer to measure its speed. The idea of flight fascinated him throughout his life, and he spent hundreds of hours watching and studying birds as they flew. How did they manage to stay up in the air? How did they soar and maneuver so magnificently? He was determined to find out.

He did not do a lot of experimentation in the way that Galileo and Newton did, but he learned a lot by observation and study. Also, unlike other scholars of the time, he had little formal education, and he never attended university. Basically, he was self-taught. Later, however, he studied under the mathematician Luca Pacioli. Early on, many of his scientific pursuits were associated with his interest in art and painting. He studied the properties of light in considerable detail. Also, for both his painting and sculpting he needed a detailed knowledge of muscle structure and anatomy.

Although he performed no experiments, he kept detailed notes of what he observed and did. They were recorded in thirteen thousand pages that included elaborate descriptions and drawings. In particular, he left a large number of drawings of his military inventions.

Worried about infringement and the possibility that some of his military inventions would end up in the wrong hands, he used a mirror-image technique for writing in his journals. It could only be understood by reading it using a mirror. In many cases he also left out critical information or changed his diagrams slightly.

He published little during his lifetime, but most of what he left appears as if it had been set up for publication. In other words, it was in the proper form for publication.

So, how much physics did Leonardo understand? A clear scientific understanding of most of the basic principles of physics had not yet been developed. The basic principles would be formulated later by Galileo, Newton, and others, but there's no doubt that Leonardo had an intuitive understanding of many of the basics. He certainly understood such things as force, mass, and inertia, and he knew the difference between accelerated and uniform motion. And many of his devices used wheels, hand cranks, and circular disks, so he had to understand angular speed and motion.

In particular, he made extensive use of simple machines that employed levers, wheels and axles, cogged gears, screws of various types, pulleys, and inclined planes. A machine is defined in physics as a device that makes work easier. Basically, a machine moves a force from the point where it is applied to another point where it is used. One of the simplest machines is a lever (a board with a fulcrum). In the diagram it's easy to see that we can raise a large weight with relative ease by applying a smaller force over a longer distance at the opposite end of the board.

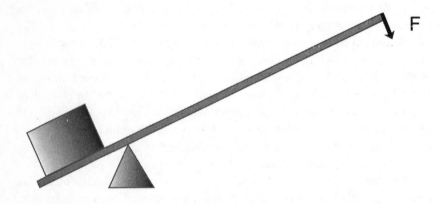

LEONARDO'S MILITARY INVENTIONS

City-states were frequently at war with one another, and it was important that they had an edge wherever and whenever possible, so military engineers and inventors were in great demand. Furthermore, there was always the danger of invasion from other countries.

Because Leonardo was employed so frequently as a military engineer, most of his inventions were war machines. Let's look at each of them in detail.[2]

Armored Tank

Leonardo proposed a design for an armored tank while he was working for Ludovico Sforza. It was a turtle-like shell that was operated by a system of gears and propelled by a crank that turned wheels. It relied on the muscle power of eight men. Guns projected out the sides in all directions, so it could have been driven into the front lines, where it would have had a devastating effect on the enemy. It was designed to be bulletproof so that the men inside would be protected from outside fire. Leonardo's diagram had a flaw, however, but there's little doubt that he added the flaw purposely.

Machine Gun

Leonardo's machine gun was not the same as our modern machine guns, which fire at a high rate through a single barrel. In Leonardo's model, eleven muskets were mounted on each of three boards in a triangular arrangement. The entire gun could be rotated so that the first layer could be fired, then left to cool. The second layer could then be fired and left to cool while the first layer was reloaded, and so on.

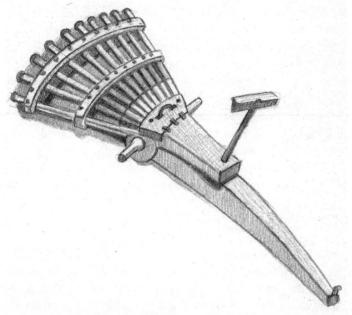

Leonardo's machine gun (several guns mounted beside one another).

Flying Machine

Leonardo's vision of a flying machine would eventually be used in war, but at the time he envisioned it, he didn't think of it as a weapon of war. As noted earlier, Leonardo was fascinated by the possibility of manned flight throughout most of his life, and he spent a great deal of time studying birds in flight. As a result, he eventually invented a device that he hoped would allow humans to soar through the air like birds. Its main feature was two wings that were operated by a crank. There is some evidence that he actually tested a model of it.

Parachute

Leonardo was not only interested in flying above the surface of the earth; he was also interested in floating down through the air to land safely after jumping from a great height. Although he certainly didn't understand gravity as we do today, he had an appreciation of it, and he did have a basic understanding of aerodynamics. We now know that when you jump from a great height, two forces are acting on you: the force of gravity is pulling you downward at 32 ft/ sec^2, and there is also an upward force due to the air you are falling through. As a result, your velocity does not increase indefinitely, as you might think. The upward force from the air slows you down until you finally reach what is called your "terminal velocity." It depends on your weight, your shape, and the air pressure. For a skydiver (diver with unopened parachute), terminal velocity is approximately 120 miles per hour. If you were to hit the ground at this speed there wouldn't be much left of you. What you want is something that slows you down before you get to the ground so that you survive, and, of course, this is where the parachute comes in. Leonardo made drawings of a pyramid-shaped framework covered with cloth that was quite similar to our modern parachute, and according to tests it likely would have worked.

As a takeoff from his parachute he also designed a glider that also likely would have worked. Again, the first gliders didn't emerge for another hundred years.

Helicopter

Closely associated with the inventions above was a simple helicopter. Leonardo got the idea from a Chinese toy that came into his possession. The device he envisioned was a giant whirling pinwheel in the shape of a screw. He knew enough about aerodynamics to know that as it spun it could produce an upward force, and as this force built up under the blade, it would lift the craft into the air as a result of the pressure produced. In his model several men on cranks were needed to turn the blade. Unfortunately, he didn't know about Newton's third law (for every action there is an equal and opposite reaction), so his idea wouldn't have worked. Nevertheless, it was an ingenious thought for the time.

Diving Suit

One of the greatest dangers to seaports or cities on large rivers was invasion from the water. You could fight ships by bombarding them using cannons on the shore, but this usually wasn't too effective. Leonardo hit upon another idea for destroying them. He made drawings of diving suits that could be used by men under water. According to his idea, they would cut holes in the bottom of ships and sink them. Divers would carry breathing hoses connected to a bell containing air they could breathe. He even designed a facemask with glass goggles that would allow the diver to see underwater. The idea was, of course, one that is commonly used today.

Giant Crossbow

Although crossbows had been used for many years, and muskets and cannons were now beginning to replace them, the crossbow was still a frightening weapon that created considerable fear. Leonardo's giant crossbow appeared to be designed mostly to scare, intimidate, and terrorize the enemy. It measured twenty-seven yards across and was carried on six wheels. It was designed to fire large stones or perhaps flaming bombs rather than arrows. A crank was used to pull back the bow to load the device.

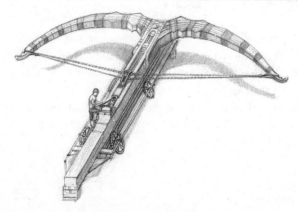

Leonardo's giant crossbow.

Water and Hydrodynamic Inventions

Water and hydrodynamics played a large role in Leonardo's inventions. As mentioned earlier, he spent years studying the motion of water as it struck various types of surfaces. This led him to design several machines that used the force of water. Paddle wheels, for example, were used in several of his devices. In addition, he designed light, movable bridges that could be assembled quickly so that troops could cross rivers.

Ball Bearings

Another invention that might not seem as important as some of the others is the ball bearing. Ball bearings are spheres that roll smoothly between moving surfaces and help reduce friction. Leonardo used them quite effectively in many of his machines in a way in which they had never been used before. And there is no doubt that they are used extensively today. Driveshafts, for example, would not be possible without them.

First Automobile and Computer

He also designed a programmable driverless vehicle that he no doubt thought of as a toy. In essence it was little more than a small cart, but it was powered by a spring similar to those used in early clocks, and it was able to move on its own when the spring was wound up. Of particular importance, a group of gears was used that forced it to travel a certain route.

Convex Lens–Grinding Machine

There is no indication that Leonardo invented any sort of telescope or microscope, but he did make convex lenses, and he didn't grind them by hand. He had a lens-grinding machine that did the work for him, and he left detailed plans for constructing it.

In his lens-grinding machine he used a handle that rotated a wheel, which, in turn, operated a gear that rotated a shaft that turned a geared dish. The glass to be ground sat in the dish.

Mortar Shells and Cannons

One of the major problems with mortar shells was their stability in flight. Leonardo showed that adding fins to a mortar increased its stability, and he made other advances in mortar technology. In a letter written to Ludovico Sforza he stated, "I also have types of mortars that are very convenient and easy to transport. . . . When a place cannot be reduced by the methods of bombards either because of its height or location, I have methods for destroying any fortress or stronghold, even if it be founded upon rock."[3]

Leonardo's notebooks also show cannons that hurled large numbers of small stones and created considerable smoke upon striking their targets. And he made drawings of triple-barreled cannons and also a steam cannon.

Other Helpful Inventions

Scaling ladders that could be used to get over castle walls were quite crude at the time. Leonardo designed ladders that could be adjusted in height and that were particularly light.

Also among his drawings were plans for double hulls on ships, which would give them considerable protection from sinking. And finally, one of his most surprising and innovative constructions was a robotic man. Amazingly, the robotic man could stand, sit, move its head, raise and lower its arms, and open and close its mouth. It was built using pulleys, weights, and gears. It was built primarily for the entertainment of Ludovico Sforza.

LEONARDO'S ATTITUDE TOWARD WAR

Since he spent so much time designing engines and machines that would be used to kill humans, it might be thought that Leonardo was fascinated by war, and perhaps that he enjoyed it. But quite the opposite is true. He actually detested war and killing of any kind—not only the killing of people, but also the killing of animals. In fact, he could not bring himself to eat animal flesh. He was a vegetarian throughout his life, and he would often buy birds that were destined for slaughter in the marketplace so that he could set them free. Although he frequently stated that he hated and felt guilty about what he was doing, designing war machines was one of the best ways for him to make a decent living.

He was no doubt thankful that most of his killing machines were never actually built in his lifetime. Indeed, he didn't even publish them, and they weren't published until 165 years after his death.

TARTAGLIA

Cannons were continually being improved, and their range was increasing, but there was still a serious problem: accuracy. Most shells went over the heads of the enemy soldiers or fell in front of them. It was understood that a cannon's range depended somehow on the angle at which it was pointed, but little else was known. The man who would solve the problem—at least partially—was Niccolò Tartaglia.[4]

Tartaglia was born in 1500 in the northern Italian town of Brescia in 1500. His father was a mail carrier who was murdered when Niccolò was only six. As a result, his family, which included his mother and a sister, was thrown into poverty. To makes things worse, Brescia was invaded by French troops a few years later, when Niccolò was twelve. To the dismay and annoyance of the powerful French army, the Brescian military held them off for seven days. And when the Brescian forces were finally overcome, the French leader was so annoyed by their fierce resistance that he decided to kill everyone in the town as revenge. Tartaglia, his mother, and his sister hid in the local cathedral, but it didn't help. A French soldier found them and slashed Niccolò across the face and jaw with his saber. Thinking he had killed him, he left. When the French left the town, Niccolò's mother took him to their home and nursed him back to health, but the wound left him with a severe scar across his face and a bad stutter. As a result, Niccolò took on the new name Tartaglia, which means stutterer.[5]

Niccolò received some education early on, but he was mostly self-taught. He soon found that he was proficient in mathematics, and concentrated on it for several years. Eventually he learned enough mathematics to become a teacher, and he got a job teaching in Verona, but he was poorly paid for this work. In 1534 he moved to Venice, where he continued teaching mathematics. He later became a professor of mathematics and eventually became famous for his mathematical talent and knowledge.

It was about this time that he was approached by one of the gunners in the Venetian army. The gunner told Tartaglia about his cannon's inaccuracy, asking him for advice. Tartaglia soon became intrigued by the problem. The first thing he determined was that the maximum range could be achieved when the barrel of the cannon was pointed at an angle of forty-five degrees from the horizontal (neglecting air friction). He then began looking into other problems. What path did the projectile trace out? What kept it moving after it left the barrel of the cannon? Gunners could not answer these questions at the time. Tartaglia had just translated the early works of Aristotle and Euclid into Italian, and he was familiar with Aristotle's ideas in relation to projectile motion. According to Aristotle, all motion of this type was straight-line motion; in other words, the projectile would move out of the cannon in a straight line and continue along this line until it ran out of what he called "impetus," then it would fall directly to earth.

Tartaglia made arrangements with the gunner to watch a test firing of cannonballs at several different angles. The gunners thought the problem might have been related to the cannon itself, or perhaps the gunpowder. But Tartaglia immediately saw that they were not to blame; the problem was more fundamental, and it arose because of misunderstandings of how and why the cannonball acted as it did after it left the barrel.

Following Aristotle, Tartaglia defined two types of motion: natural motion and violent motion. Natural motion was the motion of a free-falling body, such as a stone. He stated that all objects that were "evenly heavy"—in other words, objects that were made of dense material, such as earth, and had a generally smooth circular form, so they didn't have significant resistance to air—had natural motion. He stated that such bodies fell directly toward earth in a straight line at an accelerated rate, but he was not sure of the magnitude of the acceleration, and he had no idea what caused it.

Projectiles, on the other hand, underwent violent motion. The view at that time was that the projectile accelerated as it left the barrel. No one knew for sure whether this was the case, however, because the projectile's high speed made it invisible at this point. But it seemed reasonable. Tartaglia came to the conclusion that this wasn't true. He was sure the projectile started to lose speed the moment it left the barrel because it was no longer under the influence of the propelling force created by the expanding gases in the barrel.

His initial idea was that the projectile went through three phases. The initial phase was a straight line extending out from the direction of the gun's barrel. At some point, however, the ball would begin to lose "force," and its trajectory would become a curve. Finally, when it had lost all its "force," it would fall vertically to the earth. He published his ideas in his book *New Science* in 1537. But as he thought about it more, he realized that his ideas could not be correct, and he finally decided that the first stage of the projectile's flight was actually a slightly curved trajectory. Furthermore, he now became convinced that a certain amount of force was impressed into the body when it was projected into the air, and when this force was exhausted the violent motion of the projectile became natural motion. He added these refinements in his second book on the subject in 1546. He argued that the trajectory was a result of a struggle between the speed of the ball and the force that was pulling it toward the earth. He also argued that a body could possess both natural and violent motion at the same time in certain cases.

Based on this work, Tartaglia developed a "gunner's quadrant" to assist artillerymen in aiming their cannons. One leg of the device was inserted in the barrel of the gun, and a heavy weight showed the angle of elevation of the barrel. The gunner could then consult tables developed by Tartaglia that showed the range of the gun for various angles. These tables were used for many years. They were not highly accurate, but they were the best that was available at the time. Nevertheless, a new science had been developed, namely ballistics, and it would become increasingly important in warfare over the years.

It's interesting that Tartaglia eventually became very tortured and remorseful about his contributions to the killing of his fellow humans. He had experienced war firsthand as a youth, and he hated it. He worried about what God would think of his work. It bothered him so much that in a fit of remorse he decided to destroy all his papers and notes related to ballistics.

Soon, however, the French formed an alliance with the Ottoman Turks, and together they invaded Italy. With the possibility of war coming home to him again, he relented and reconstructed all his previous work, giving it to Italy's defensive forces.

Tartaglia's contribution was significant, but many questions remained unanswered. He didn't know what kind of a "force" pulled the projectile back to earth, or how strong it was, and he knew nothing about what we now call inertia. The task of developing an understanding of these concepts was left to Galileo.

GALILEO

Tartaglia made important advances in the understanding of the trajectories of projectiles, but many problems remained. A better understanding of the nature of motion was needed, along with some sort of understanding of gravity. The first real breakthrough came from Galileo Galilei in Italy. He is frequently referred to as the father of modern physics, and there's no doubt that his achievements were phenomenal. Indeed, he was the first scientist to make a significant break with the teachings of Aristotle, which had been accepted for centuries. All the problems could not be solved, but the stage was set for another of the early giants of science: Isaac Newton.

Born in Pisa, Italy, Galileo was the oldest of seven children. His father was a musician and composer who dabbled in mathematics and experimentation. He even made an important breakthrough in physics, showing that in a stretched string, the pitch or frequency varies as the square root of the tension. Galileo no doubt got his skepticism of established authority from his father.[6]

Galileo's father was well aware of the fact that musicians and mathematicians were among the lowest paid professions. He encouraged his son to go into medicine, which was highly paid. And indeed, at age seventeen Galileo entered the University of Pisa to study medicine. But he soon got bored. A class in mathematics led to an interest in math and science, and Galileo decided he wanted to switch his focus. His father was extremely disappointed, for he knew that mathematicians made no more money than musicians, but he finally agreed.[7]

Galileo lectured at the University of Pisa in 1589 and was appointed to a chair in mathematics in 1592. He then moved on to the University of Padua, where he remained until 1610.

The Ballistics Problem

Within a few years Galileo made major contributions to the study of the trajectory of projectiles, which was of critical concern to gunners. It all started with his interest in gravity. Aristotle had said that all objects fall toward the earth with a speed that depends on their weight, and for years this appeared to be reasonable. It could easily be seen, for example, that very light feathers fell much slower than heavy stones. Galileo was skeptical, and according to legend he carried several balls of different weights to the top of the tower of Pisa and released them. The balls all struck the ground at the same time. Aristotle was wrong. Actually, there is no evidence that Galileo performed this experiment, but it is an interesting story nevertheless.[8]

Galileo wanted to go further, however. It was now obvious to him that the balls accelerated as they fell, so they had different velocities at different positions above the earth. Indeed, the farther they fell, the greater their velocity, and Galileo wanted to measure their acceleration. But because objects fell so fast, it was difficult to set up a straightforward experiment. So he decided to slow things down. The best way to do this was to let the object roll down an incline. The object would speed up in the same way because gravity was still acting on it. Again, he noticed that the acceleration of the balls down the incline was independent of their mass. In other words, they all got to the bottom with the same speed, regardless of how much they weighed. A detailed study of this motion led to several important conclusions.

Earlier Galileo had discovered something similar with pendulums. While at church he had noticed objects at the end of long ropes swinging as a result of air currents in the church. Clocks were not available at this time so he used his pulse to time them, and he noticed that regardless of the distance they swung (called the amplitude) they took the same time to complete a swing. Again, it was gravity that was pulling the weight downward (along with the current), causing the objects to swing. Galileo was never able to measure the acceleration of gravity, but we now know that it is 32 ft/sec^2, and we know that it acts on all objects on earth. He did, however, show that the square of the period of the pendulum varied directly with the length of the pendulum.

Building on his discoveries in relation to gravity, Galileo decided to look at projectile motion carefully in an effort to thoroughly understand it. He imagined first of all that there was no air resistance, as he knew that the air around a projectile acted on it to change its motion. As a first step, it was best to ignore it. Second, he considered the forces that were acting on the projectile. Obviously the first force was the expanding gas from the gunpowder that thrust the projectile from the cannon barrel. Once it was out of the barrel, this force was gone, and the projectile would have a constant velocity if no other forces were acting on it. In stating this, Galileo was imagining a new concept that we now call inertia. All bodies in motion have a certain amount of inertia, and as a result of it they will continue in motion with the same velocity unless this inertia is overcome by an outside force. In the above case there was an outside force acting on the projectile after it left the barrel, namely gravity, and gravity would cause it to fall in the same way any object falls when released. The only difference in this case was that the projectile also had a horizontal velocity.

These results helped to give Galileo a better understanding of projectile motion. As a result of his observations he came to the following conclusions:

• Bodies fall with uniform acceleration (as long as the resistance of the medium is neglected).
• Objects in motion retain their motion unless some sort of force acts on them.
• The law of acceleration: the total distance from rest under acceleration is proportional to time squared.

His big break with previous ideas was the statement that a force was present only during the acceleration of the projectile. Once the force was taken away, the object no longer accelerated, but it continued at a constant velocity unless acted upon by another force. This was in conflict with Aristotle's idea that a projectile in motion was under a constant force; in other words, it had a "reservoir" of force that was gradually used up. Galileo said this was incorrect.

Galileo decided that the most logical curve the projectile would undergo as a result of this was a parabola. What is a parabola? The best way to understand it is to think of a cone (see figure). If you slice through it parallel to the base, you'll get a circle, but if you slice through it at an angle, you'll get a parabola (as long as you don't pass through the base).

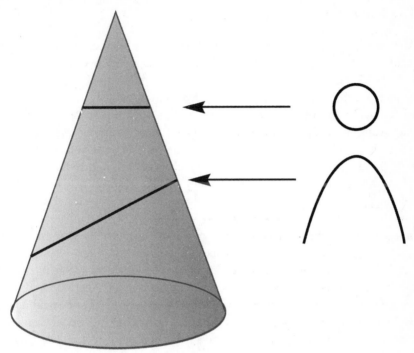

A parabola, the second curve from the top.

Military Compass

As a result of his work on projectile motion, Galileo developed a "geometric military compass." It was a takeoff on Tartaglia's device for gunners, but it had many improvements. It gave gunners a new and safer way for aiming their cannons more accurately. Furthermore, it had scales and numbers on it that told them how much gunpowder was needed for cannonballs of various weights and sizes.

Like Tartaglia, Galileo abhorred war and felt guilty about developing war weapons, but he felt they were necessary. Also, his wages were relatively low, and it was something that paid well. So not only did he develop the gunner's compass; he had over a hundred built and sold, and he made an excellent profit on them. Furthermore, he taught a class to gunners on how to use the new device, and he also wrote a book on it, which he sold.

Interestingly, with minimal modifications, the same device was eventually used in surveying.

The Telescope

Although it isn't exclusively an instrument of war, the telescope is invaluable in relation to it. The first telescope was built by Hans Lippershey of the Netherlands in 1604. Galileo heard of it within a short time, and he set out to build one. He was already familiar with the grinding of lenses, so he was able to construct one quite quickly. He completed his first telescope in 1609, and it was a significant improvement on Lippershey's model; it had a power of about three (in other words, it magnified three times). Soon after he built an improved one of about eight power, and he presented it to the lawmakers of Venice in 1609. They were thoroughly awed and impressed, and they quickly realized that it would be helpful in the event of an attack, particularly from sea. The sail of an enemy ship, for example, could be seen at least two hours before it could be seen with the naked eye, and this would give a tremendous advantage. He was awarded a stipend for building additional telescopes.

But the military use was of little interest to Galileo; he was more interested in what telescopes would show him above the earth in the night sky. And over the next few years Galileo revolutionized astronomy. He discovered that Jupiter had four tiny moons, and that Venus presented phases like our moon when seen close up. He also noticed that Saturn had a strange ring around it, and he went on to study our moon, noting that it was covered with craters. And even the sun was different than thought: it was not the pure, clear disk that everyone had assumed. It was covered with dark spots—what are now called sunspots. And finally he looked in detail at the Milky Way and found that it was composed of thousands (perhaps millions) of individual stars. Indeed, over a relatively short period of time he made more discoveries in astronomy than had ever been made in the centuries before him, and even after him.

And Galileo didn't stop with the telescope; he also constructed a microscope. Again it wasn't the first, but it was likely the best available at that time. He used it for examining insects and various other small objects.

Other Inventions

The telescope and microscope were not the only devices Galileo constructed. In 1593 he built one of the first thermometers. It was based on the expansion and contraction of air in a ball that moved water in an attached tube. He even tried to market it, but he was unsuccessful.

Galileo was one of the first to understand the role of frequency (or pitch) in relation to sound, and he made an attempt to determine the speed of light, but was unsuccessful. And he invented a device for determining how much heavier metal was than water.

Galileo is perhaps best known for his objection to the idea of the earth-centered universe that was accepted in his day. He was sure that the sun was at the center of the solar system, and he was eventually condemned by the church for his ideas.

7

FROM EARLY GUNS TO TOTAL DESTRUCTION AND DISCOVERY

For several decades after the death of Galileo there was almost continuous warfare. This period included the Thirty Years' War of 1618 to 1648, which was one of the most costly wars in terms of human life in the history of the world. The deadliness of this conflict was largely a result of the new weapons that were devised. So let's start with these weapons, and the guns, in particular.

THE GUNS OF WAR

In the previous chapters we saw how the cannon was developed and how it evolved, but within a short time after its first use men were beginning to think about something smaller that could be handheld, and soon the first hand cannons appeared. They came about mainly as a result of the problems of steel armor; that is, it could still withstand most of the arrows from the longbows (unless they happened to strike the right place) and it was quite effective against the bolts of the crossbow. Something was needed that could easily penetrate this armor. Cannon shells were certainly adequate, but they were large and unwieldy. Something smaller was needed, and it finally came in the form of the hand cannon. Hand cannons were first used in China in the thirteenth century, but they were generally inaccurate and difficult to use; nevertheless, the bullets from them could penetrate most types of armor at close range.[1]

The barrels of the earliest guns were about four feet long, and they were made from wrought iron or bronze. Attached to the barrel was a wooden stock. One of the main difficulties in using these early guns was that two men were needed because it took two hands to aim and hold them steady and another pair of hands to light and hold a match to the touchhole. It was possible for the gunner to prop his gun in a support and light his own gun, but it was difficult. The earliest hand cannons were also relatively heavy, at about twenty to twenty-five pounds, but they could fire projectiles up to one hundred yards.

Strangely, while they were relatively inaccurate, their flash and loud roar usually had a strong psychological effect on the enemy—particularly if the enemy had never seen them before. In many cases enemy soldiers fled in terror. Hand cannons were used extensively throughout Europe and Asia until about the 1520s. But as new developments in powder emerged, such as granulated powder, handguns began to improve. The first to appear after the hand cannon was the arquebus, which means "hook gun" in Dutch. What the hook referred to is still uncertain. Most believe it was the hook-shaped wooden stock. Later guns did, however, have a hook mechanism that held the match.

There is a problem with terminology, as some of the later guns were also called arquebuses. At any rate, it was first used in about 1458, and it was commonly used until about 1490. Again, it was a short-range weapon that was difficult to reload, but looked a lot more like our modern rifle than did the early hand cannon. Furthermore, advances in gunpowder had made it much more powerful. But it was heavy and usually had to be rested on a balance.

The arquebus was followed by the musket, but again there's a problem with terminology. Later on, almost all hand-held guns were called muskets, and the gunners that shot them were known as musketeers. They were muzzle-loaded and had a smooth-bore barrel. The earliest handheld guns were usually held against the chest, but within a short time they were designed for the shoulder, and gunsmiths eventually design curved stocks so that the gun could be held up against the shoulder to stop the recoil. This represented a considerable improvement over the arquebus in that a musketeer could usually get off two shots in about three minutes.

Over time the early musket evolved into the matchlock musket. Its main advantage was that it got rid of the problem of igniting the primer using a hand-held match. The match was now attached to the gun and was applied when the trigger was pulled. Eventually the match became a slow-burning fuse, or, more exactly, a smoldering piece of cord. The matchlock also now had a primer pan. A spring-loaded lever was attached to the metal hook that held the smoldering cord. When the shooter pressed the lever with his fingers, the "match" was lowered into the priming pan, which was filled with powder. The priming pan was attached to the touchhole, which led to the charge in the barrel. The flash in the priming pan ignited the powder in the touchhole, which, in turn, ignited the powder in the barrel. Within a short time, however, the lever was replaced with a trigger.[2]

The matchlock was the main military weapon for many years, but it was cumbersome to load and fire. Before firing, the gunner had to go through many steps:

- Pour powder down the barrel, then place a wad and a bullet on top of it
- Fill the primer with different powder from another flask
- Cock the hook holding the smoldering cord, then blow on it and make sure it would ignite powder
- Pull the trigger

And even with all this, much of the time it wouldn't fire. If it was raining or the weather was bad the powder would be ineffective. Furthermore, the gunner was in danger because he carried so much open powder and the fuse was always lit. Accidents were common; sometimes, in fact, the gun would explode in his hands. Nevertheless, the matchlock was used for many years, and over the years it was improved. The length of the barrel went from about four feet down to about three. Rests or balances for the gun eventually became obsolete, and the powder gradually improved.

A significant improvement over the matchlock came in the early 1500s, but for the most part it was rarely used in military guns, and main reason was that it was expensive. The new gun became known as the wheel lock. The lighted fuse was the main problem: it was useless in rain, and it could be easily seen by the enemy. What was needed was a mechanism that created a spark that could light the primer pan. The design for such a mechanism was first discovered by Leonardo da Vinci in about 1490, but he kept most of his inventions secret, so it's not known if it helped the military at this time. The same design was found in a German book in 1507 in Austria, and it was eventually built by German gun makers early in the 1500s.[3]

The mechanism is similar to a modern cigarette lighter. In short, the spark was generated by a spinning grooved steel wheel that was pressed against a piece of pyrite. Another critical development was that the flash pan now had a cover to keep the powder in it dry. In preparing to fire, the gunner slid open the flash pan and poured powder into it and then slid the lid shut. The steel wheel was in the flash pan, and the lever holding the pyrite was above the pan, held in place by a spring. When the trigger was pulled, the wheel began spinning, the lid of the flash pan slid back, and the pyrite slammed down into the spinning wheel producing intense sparks. The sparks lit the primer, which in turn lit the powder in the touchhole, and it triggered the explosion in the barrel.[4]

German gun makers were quite enthusiastic about the wheel lock, but its mechanism was expensive and delicate. Basically, it was too expensive to be mass-produced for the military. However, it was used by aristocrats as a hunting weapon, and later the same mechanism was used in pistols. Pistols, in fact, were now favored by the cavalry, as they were easy to hold and fire.

THE WAR AT SEA

Handheld guns spread quickly throughout Europe, and soon the long bow and the crossbow were gone. The musket was the main weapon, and its effectiveness tended to accelerate warfare in an already fragile arena. And while things were accelerating on land, developments were being made at sea. For the most part, though, progress was slower here, and there were more problems. One of the biggest problems was the mounting of large cannons on the deck of a ship. They were heavy, and if too many of them were mounted they would make the ship unstable.

But this wasn't the most serious problem. Navigation was a hit-or-miss process, particularly on the open sea. Because of this, most sea captains preferred to stay within sight of land. But they could only do this so long. It soon became known that the region beyond the open seas was rich with treasure. Not only was there gold, but there was sugar, tea, and spices that could be sold at a tremendous profit. The sea, however, now held another danger: pirates. Small, fast ships were waiting for large ships laden with treasure, and they quickly attacked any they found.

England, Spain, France, and Portugal were the main powers at the time, and they all wanted to build up their navies, both because of the lucrative trade with Asia and other lands and also because they needed a strong military force at sea. One of the first to realize a strong navy was critical to survival was Prince Henry of Portugal. He was born in 1394, the third son of King John I. At the age of twenty-one he headed up a military force that attacked and captured the military outpost of Ceota, in the Straits of Gibraltar.[5]

Because the feat so impressed his father, Henry was allowed to set up a naval institute in the southwest of Portugal at Sagres. Henry, the navigator, as he eventually became known, knew that a powerful navy was critical to the success of a country, and he set out to build the best navy in the world. It was also the key to incredible treasures beyond the seas.

As a first step, Henry began a search for the best mathematicians, astronomers, cartographers, and geographers in Europe. And in 1418 he assembled them all at his naval institute. He had several goals for the institute. One of the most important was developing better navigation, but he also wanted to design faster and better ships. Basically, his institute was a research and development facility that included one of the best libraries in the world.

Henry soon had two major goals: locate the best naval route to the main trade areas of Asia, and explore the west coast of Africa. The continent of Africa was a great unknown at the time, but there was a lot of speculation about the treasures that might be found there.

One of the first things that Henry did was develop and build a new type of ship, which was faster and more maneuverable than most ships on the seas. It was called the caravel. And he quickly sent out the first caravels to explore the west coast of Africa. In particular he wanted to explore as far south as possible. (Although he sent out many expeditions, he was not on any of them himself.) But there were problems. One of them was that navigators needed Polaris (the North Star) to navigate, and to their surprise and frustration, Polaris disappeared over the horizon when they moved too far south. Furthermore, there were many tales of monsters, wild waters, storms, and so on in the region, so sailors were very cautious about going too far.

Navigators had compasses to help guide them. The device, which consisted of a sliver or needle of magnetite, balanced so it could spin freely, had been developed many years earlier by the Chinese. They also had sailing charts, but they were inaccurate, as little was known about the region beyond the open seas.

Surprisingly, Henry could have made a major discovery in navigation on the open seas, but he let it slip through his fingers. An Italian cartographer and mathematician, Toscanelli, was working on new charts of the world, as it was known at the time. Born in Florence, Toscanelli had been educated in mathematics at the University of Padua in 1424. His main interest in his early years was astronomy, and he made numerous observations of comets, but he eventually became interested in cosmography—the study of the overall earth, as it was known. What did the overall earth look like? He was acquainted with the maps of early Greek geographers, and he had knowledge of Ptolemy's work and Marco Polo's trips to Asia. Using these sources, he set up a map showing Europe and Asia; he was convinced that their land mass covered approximately two-thirds of the surface area of the earth. He overlaid his map with a grid (squares of a particular size that covered the entire map). Not only did they cover the land areas, but they also covered the open sea. Excited about his new maps, he took them to Henry's naval institute, sure that his scholars would be excited about them also. To his disappointment, they showed little interest.[6]

Undeterred, Toscanelli then took his maps to Spain, where they were enthusiastically accepted. Indeed, it was just before Columbus was setting out on his well-known voyage of 1492 across the open sea, which resulted in his discovery of America. Columbus was delighted with the new maps and was said to have used them on his trip.

HENRY VIII OF ENGLAND

When most people hear the name Henry VIII they think of his problems with his many wives. Few know that he developed one of the strongest navies in the world at the time. He increased the English navy from eight ships to forty-six warships and thirteen other smaller vessels. And he was also responsible for important advances in the physics of naval warfare. He certainly wasn't interested in science, but he was determined to strengthen the English naval forces, mostly because he needed to protect English sea trade, which meant riches. Gold, silver, sugar, spices, and tea enticed him. But ships laden down with gold and other treasures were bait for pirates. Furthermore, Spain, Portugal, and France were also looking at what the lucrative trade market could do for them. Henry was also worried about an invasion from France.[7]

He had to make his ships larger so that they could carry more, and at the same time he had to arm them. His shipbuilders, however, warned him of the problem of stability with large, heavy cannons on the deck. The major problem was the ship's center of gravity, or center of mass. The center of mass is basically the balance point of the ship; in short, it's the point that has an equal mass in all directions around it. In two-dimensional objects the center of mass can easily be determined by finding the point where the rotational force in one direction (clockwise) around the point is equal to the rotational force in the opposite direction (counterclockwise). Although it is simple and relatively easy to find in two dimensions, determining it in three dimensions can be complicated, and a ship is, of course, a three-dimensional object.

Basically, if a ship is to remain as stable as possible in water, its center of mass must be as low as possible in the water. This means that most of the weight has to be below sea level. Putting heavy guns on the deck would raise the center of mass. The only solution was to put the guns below the deck. But how was this possible? Henry and his shipbuilders decided that holes would have to be built in the sides of the ship. When not in use they would have to be covered with waterproof covers. These holes became known as gun ports.

In addition, however, there was the problem of recoil. If the guns were large and heavy, and there were very many of them, the recoil could lead to overall instability of the ship. Henry decided to put the cannons on wheels and leave sufficient space behind them for the recoil.

Over the years Henry built many large gunships, but his favorite was the *Mary Rose*, named for his sister, Mary Tudor. It was built between 1509 and 1511. It was a six-hundred-ton ship, the second largest in his navy, and it was designed for close-up fighting, with fifteen large bronze guns, twenty-four smaller iron guns, and fifty-two antipersonnel guns. She was designed to sail into the enemy with guns blazing, then turn broadside and shoot all her broadside guns at it, and finally maneuver around so she could shoot all her guns from the other side.

In July 1545, however, the *Mary Rose* led the English fleet into battle against an advancing French fleet. Faster than the rest of her fleet, she met the French with all her frontal guns blazing, and then turned so her broadside guns could be used. Suddenly, however, a gust of wind caught her while her lower gun ports were open, and water poured into them so quickly she sank and took most of her crew with her. In 1545, however, Henry went on to build another, even bigger, ship called the *Great Henry*.

But as large and powerful as his ships were, they still had a problem: safe navigation on open seas.

WILLIAM GILBERT

The compass was the major navigational instrument of ship navigators. Normally the compass needle pointed north at the Pole Star, Polaris, but at sea compasses were not reliable. Sometimes the needle pointed directly at Polaris, but at other times it deviated considerably from it, and there didn't appear to be a consistent reason for the deviation. What was wrong? The problem was presented to William Gilbert, a prominent physician in London.

Gilbert was born in 1544 and educated at St. John's College, Cambridge. He obtained his medical degree in 1569 and left to practice medicine in London. When the compass problem was presented to him he knew almost nothing about magnetism or physics, in general. Furthermore, very little about magnetism or the "amber effect," as it was called, was understood at the time. Some idea of the understanding of magnetism at the time can be gleaned from the commonplace contemporary idea that assumed that garlic affected a magnetic field. Gilbert showed that it had no effect.[8]

When Gilbert started his work, which consisted of both research and experimentation, the basic properties of loadstone (magnetic iron ore) were generally known, the amber effect (amber could acquire a charge when rubbed) was known, and the compass had been used by navigators for years. He soon showed that the "electric" field associated with amber was not the same thing as the field associated with magnetism. He did this by showing that amber's electrical effect disappeared with heat whereas a magnetic field did not. We now know, of course, that a magnetic field does change when severe heat is applied.[9]

Gilbert also invented the versorium, which consisted of a freely supported metal needle and a round loadstone. Its needle moved in response to either an electrical or magnetic field, so it was similar to our present-day electroscope in which two small metal "sheets" repel one another when given the same electrical charge. As a result of his studies with the instrument, Gilbert suggested, correctly, that the earth was a giant magnet. And it was the field of the earth that influenced the compass. In particular, he showed that the polarity of the earth was similar to that of a loadstone magnet. Up to that time most people believed that Polaris somehow attracted the compass needle. He went on to argue that the center of the earth was composed of iron, and, of course, we now know that it is. Another important property of magnetism that he discovered was that if a magnet was cut in half, the two halves both have a north and a south pole. In other words, they form new and completely similar, smaller magnets.

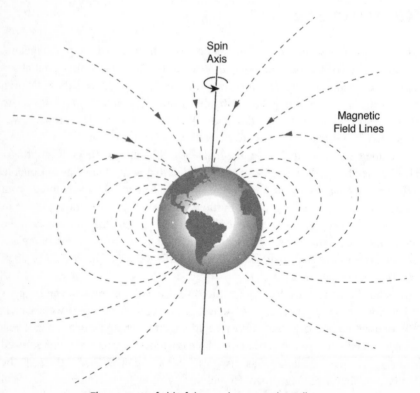

The magnetic field of the earth, as seen by Gilbert.

It was also believed at the time that the stars were on a fixed sphere that rotated around the earth. Gilbert suggested that it was the earth that rotated, not the stars. Furthermore, he didn't believe the stars were on a fixed sphere. And indeed, it was true that the earth rotated, and it was now clear why navigators sometimes had problems with their compasses. As Gilbert pointed out, the axis of magnetic earth didn't line up perfectly with the axis of its spin. In effect, there was a true north and a magnetic north, and the compass always pointed at magnetic north while Polaris was only approximately lined up with true north. The biggest problem was that the deviation, or apparent distance between the two poles, was different at different points on the earth. There was soon hope that tables could eventually be made to help navigators compensate for this. Several people worked on the problem, but early navigators were never able to use the tables effectively. However, at least they finally knew why compass needles acted up.

Gilbert gathered all his results into a book he titled *De Magnete*, which he published in 1600. For many years it was a standard work on electricity and magnetism. As a result, Gilbert became quite famous: he was elected president of the Royal College of Physicians, and in 1601 he became personal physician to Queen Elizabeth I. Much of the terminology now used in electricity and magnetism is, in fact, due to Gilbert: electricity, electrical force, electrical attraction, magnetic poles. Because of this, Gilbert is sometimes referred to as the father of electricity and magnetism.

We now know that if amber is rubbed with fur it generates what we call a negative charge; similarly, when a glass rod is rubbed with a silk cloth it accumulates a positive charge. Electrical field lines are assumed to point away from a positive charge and toward a negative charge. Also, there is a force between two charges: a repulsive force between like charges and an attractive force between dissimilar charges. And just as an electrical charge is surrounded by an electric field, a magnet is surrounded by a magnetic field, with the field lines assumed to be pointing out of the north pole and into the south pole. Again, similar poles repel and opposite poles attract.

Few, if any, weapons of war were devised using Gilbert's discoveries in the years soon after he made them, but we now know that his work is the basis of all our knowledge of electromagnetism, including electrical currents and circuits, and, as a result, most of the devices in our modern world are based on Gilbert's work. So Gilbert's discoveries eventually made a large contribution to warfare.

THE PROBLEM OF LONGITUDE

Gilbert had set the stage in showing that there were two norths: true north and magnetic north. And for years, scientists, astronomers, and others worked to see if this difference could be used to make navigation at sea safer. The English astronomer Edward Halley took several voyages to study the magnetic variance between the two. He even developed maps as the basis of a new method, but the problems seemed insurmountable.

For navigation, knowledge of both latitude and longitude was needed. The astronomer Eratosthenes had proposed many years earlier (3 BCE) that positions on the surface of the earth could be located by a grid of crossed lines similar to our present lines of latitude and longitude. Hipparchus took the idea a step further in 2 BCE; indeed, he even proposed that one of the series of lines (longitude) should be associated with time.

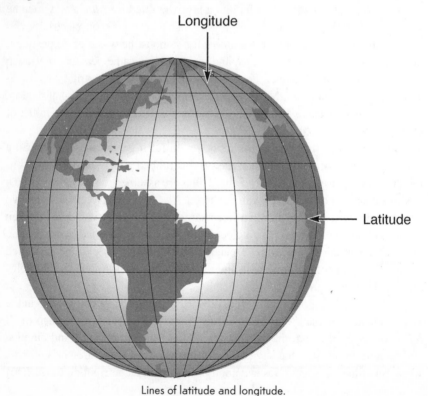

Lines of latitude and longitude.

Latitude was relatively easy to determine, both on land and at sea. It could be found during the day by measuring the altitude of the sun above the horizon at noon and comparing it to a prepared table. The problem was longitude; it was relatively easy to measure on land, but it was difficult to determine at sea, mostly because time was involved. At this time clocks depended on a pendulum, which worked great on land, but pendulums were very unreliable when set on the rocking and rolling deck of a ship.[10]

Because of this problem, ship captains would frequently ignore longitude and sail to the latitude of their destination, then follow it the rest of the way to their destination. But this was time consuming because it was not the shortest route, and it could be dangerous. Many shipwrecks resulted because of it. The problem became so serious that a prize of millions of dollars in today's money was offered in England to anyone who could come up with a solution. It was a great incentive. Other countries, including France, Spain, and Holland, soon offered similar prizes.

It was known at the time that the earth rotated three hundred sixty degrees on its axis in a day, or fifteen degrees an hour. At the equator this amounted to sixty miles every four minutes, and because of this, noon occurred four minutes later for every sixty miles of travel west. This told navigators that they could determine their distance from their home port if they knew the exact time at home and the exact time aboard the ship so that they could compare the two. Tables of latitude and longitude would have to be used in conjunction with this method, and they had already been set up. Determining the difference between the two clocks accurately, however, was not easy because the clock aboard the ship was so inaccurate. A better method was needed for determining the time on the ship and also the time of a distant reference point while on the ship.

Navigators eventually turned to astronomers. The moon and stars were, in a sense, very accurate clocks—particularly the moon as it moved past the stars. Its movement had been tracked very accurately for years, and the times at which it would eclipse various stars were well known. Because of the huge prize, observatories were set up in both England and France. The British Royal Observatory was located at Greenwich, and the French observatory was located at Paris, and soon a serious problem developed. Each of the observatories used their own zero meridian (origin of latitude).

The moon's apparent flight through the stars seemed to be an excellent clock. The moon moved 360° around the sky roughly every 27.3 solar days, or 13 degrees per day. Halley proposed using a telescope to observe the time when the moon occulted (eclipsed) various stars in its flight. He even prepared elaborate tables, but the method didn't work out well because few bright stars were in the path of the moon.

Other astronomers tried other methods. Even Galileo had suggested that the orbital period of Jupiter's four bright moons could be used as a clock. But they were difficult to identify for an observer at sea, particularly as the sea rolled beneath the observer.

In the end, the best solution was an accurate clock for the ship, and it came when the British clockmaker, John Harrison, realize that pendulums could not be used for clocks at sea. He devised a spring-driven clock, and it worked beautifully.

THE THIRTY YEARS' WAR

As cannons and handguns improved, they became more deadly, and, as a result, things soon began to change. For the most part, battles became less intimate, with one group firing at another from a distance. Gunners never saw the people they killed. At the same time, however, warfare became more devastating and gruesome. Armies would march through villages pillaging, burning, looting, and raping, and sometimes they would even obliterate entire villages.[11]

Muskets were by this time fairly accurate at several hundred yards, and cannon bombardment had become devastating. There was little protection from either of these. Because of this, warfare became common, with few breaks between battles. One of the most devastating wars was the Thirty Years' War, which lasted from 1618 to 1648. It actually wiped out a large fraction of the male population of several countries, and it was one of the longest continuous wars in the history of the world. It began as a religious war, mostly between the Protestants of France, Sweden, and Holland and the Catholics of the Holy Roman Empire. At the time the Holy Roman Empire was made up of a collection of largely independent states, including Spain, Austria, and Bavaria. As time passed, however, religion began to play less and less a role, and the war developed into a more general conflict based on politics.[12]

Many of the main battles of the Thirty Years' War took place in what is now Germany. Needless to say, developments in physics and all sciences were at a standstill during these years. By the war's end, most of the countries involved were not only bankrupt, but also in ruins. Physics did play a role, however, in its contribution to the weapons and tactics used.

The war began in 1617 when an Austrian prince, Ferdinand II, was chosen to become the king of Bohemia. Within a year of ascending to the throne Ferdinand, a Catholic, began closing Protestant churches. The Protestants, as expected, rebelled, and when Ferdinand sent two Catholic counselors to the

Hradčany Castle in Prague as administrators of his new government, Protestants seize them and threw them through a plateglass window that was seventy feet above the ground. Miraculously they survived, but the war was now on.

The Catholic king of Spain and the monarch of Bavaria sided with Ferdinand. Several German princes sided with the Protestants. Between 1618 and 1625 the Protestants suffered defeat after defeat. The key players on the Catholic side where the Habsburgs, a royal family that ruled the Holy Roman Empire, along with Austria, Bohemia, and Spain. Many of the other countries in the region feared the Habsburgs, and the French, English, and Dutch formed a league to oppose them. They backed Christian IV of Denmark and encouraged him to invade Germany in support of the Protestants. Against Christian's army was a Habsburg army under Albert of Wallenstein, and Christian's army was crushingly defeated. Over several years the Protestants lost considerable land and suffered defeat in most battles. Then came Swedish intervention.

SWEDISH INTERVENTION

The Holy Roman Empire and its Catholic allies were soon in for a surprise. The young king of Sweden, Gustav Adolphus, began to worry that the war was getting too close to Sweden; he feared a move against Sweden and decided to act before it happened. Adolphus had become king of Sweden in 1611 at the age of seventeen. He was young, but he had been prepared well by his father, and he was a natural leader. Over the years he had taught himself the latest techniques in artillery, military strategy, logistics, and organization. Furthermore, he already had considerable experience leading troops to war; at the age of thirty-one he had led Sweden to war against Poland, and won.

His troops were some of the best in the world. Gustav drilled them continuously, and he was a strong disciplinarian. He wanted perfection, or close to it; in particular, he wanted maximum gunfire from his troops at all times. In his effort to achieve this goal he reduced the weight of muskets so that they were easier to handle, and he introduced paper cartridges and containers of premeasured powder. Loading had to be done as quickly as possible. His military could fire with high accuracy at speeds three times faster than most of his rivals. In addition, he made important tactical changes. His troops charged the enemy from the front and sides, firing as they went. Then they would quickly retreat and reload for another assault. He emphasized attack over defense, and mobility was also emphasized. And finally, unlike most armies, his army was cross-trained. Infantry and cannon gunners could easily exchange places, and everyone was taught to ride horses, so a cavalry man could easily be replaced. He made so many advances, in fact, that he is frequently referred to as the father of modern warfare.[13]

So when he attacked northern Germany in 1630 he was well-prepared, and he easily overcame the opposing Catholic army. Unlike most armies that plundered and looted the countries they occupied, Gustav did not allow his troops to plunder and loot. Forging forward, Adolphus met the army of the Count of Tilly in 1631, and he quickly overcame it. Then he continued across Germany to the Rhine, where he stopped in preparation for an invasion of the Holy Roman Empire. The following year he invaded Bavaria again, quickly overcoming his Catholic opponents.

In 1632 he had a coalition army of about twenty thousand men. General Wallenstein, whom he had fought earlier, led a Catholic army that had almost the same number of men. They approached one another at Lützen, Germany. Adolphus and his troops bedded down and prepared to attack at dawn, but when they woke, the entire area was immersed in a thick log. The delay helped Wallenstein get his cavalry into position, and it caused several problems for Adolphus. The fog didn't lift, and Adolphus finally decided to attack under the cover of fog. But when he attacked, confusion reigned; it was difficult to distinguish the enemy, and Gustav and a small contingent of mounted soldiers soon lost contact with the main branch of their cavalry. The confusion that followed led to a tremendous slaughter on both sides, and during it, Adolphus was hit by a bullet. His horse panicked and began running wildly through the fog until Adolphus finally fell off. As he lay on the ground several enemy soldiers who were probably ignorant of who he was shot him again. In the end the Swedes won, but their leader was now dead, and this was a tremendous blow to them.

Interestingly, Wallenstein survived the war but was assassinated shortly after it. And the overall war was still not finished. It continued for another sixteen years, coming to an end in 1648. Adolphus was hailed as a hero in Sweden, and he has been revered ever since. He is now referred to as Adolphus the Great.

A NEW ERA OF DISCOVERY: ISAAC NEWTON

Few advances in physics occurred in Europe during the Thirty Years' war, and England was still weak, so little happened there. But soon one of the greatest periods of scientific productivity would come, and one man, Isaac Newton, was mainly responsible for it. Newton had almost no interest in the military, and he did not work directly on any military project, but his discoveries had a tremendous effect on weapons and warfare, and because of Newton's insights, humankind for the first time had a fundamental understanding of the physics behind them.

Newton was born in January 1643 at Woolsthorpe Manor in England—the same year that Galileo died. His father was a relatively well-to-do farmer, but he died before Newton was born. Newton's mother remarried shortly after his birth and left him with his grandparents. Later, he attended boarding school in Grantham, and from age twelve to seventeen he attended Kings School in Grantham. When he graduated, his mother decided to make a farmer out of him, but it didn't work out. He had no interest in farming, and the master at Kings School finally persuaded her to let him attend Cambridge University.[14]

Isaac Newton.

In June 1661 he entered Trinity College at Cambridge to study mathematics, physics, astronomy, and optics. Not much is known about his years at Cambridge, but his outstanding ability came to the attention of one of his teachers, Isaac Barrow. In 1665 the bubonic plague struck; Newton had to return home. The year that he spent there is now considered to be one of the most important years in the history of physics. It was during that year that (according to legend) he saw an apple fall from a tree in his yard and began to wonder why it fell. The event soon led to his famous "law of gravitation"—one of the most important breakthroughs in physics. During the same year he is said to have invented calculus, but strangely, he kept it secret for many years. And in the meantime a slightly different form of calculus was discovered by Leibniz in Germany, and, as a result, there was considerable controversy later about who actually invented calculus.

Soon after returning to Cambridge he was appointed professor of mathematics, a position that he held for the rest of his life. And during his early years he continued to make fundamental discoveries, particularly in the field of light and optics, but also in relation to motion and dynamics. He presented some of his discoveries to scholars at Cambridge and was surprised that he was criticized quite severely—and Newton was not a man who could take criticism. He continued his experiments, and he continued to make important discoveries, but he filed them away and kept them to himself for years. Indeed, if it hadn't been for the astronomer Edmond Halley, he might have taken them to his grave.

In 1687 Halley and a friend, physicist Robert Hooke, were having an argument about the mathematical form of the law of gravity; they had ideas about it, but they were uncertain. Halley knew Newton, and he was sure that Newton could resolve the argument, so he went to him. And sure enough, Newton had the answer. He told Halley that he had proved mathematically that it was an inverse-square law, and he offered to show it to them. He searched for the calculation he had made, but he couldn't find it, so he promised to send it to Halley later. Halley received it a few days later and was amazed. He went back to Newton to discuss the results with him and was amazed even more that not only had Newton discovered the law of gravity, but he had made numerous other discoveries that he had never published. This eventually led to one of the most important books ever published in physics, called *Philosophiae Naturalis Principia Mathematica*, or, as it's more commonly known, the *Principia*.

Contained within the *Principia* were the three basic laws of motion, now referred to as Newton's laws. The first law stated that *a body will continue in a state of rest or motion in a straight line, unless acted upon by a force*. At the time this seemed to defy common sense. It didn't seem possible that objects in uniform motion continued their motion indefinitely. But unless a force acted on them to change their motion, they did. Newton's second law was concerned with this force. It stated that *an acceleration produced by a force acting on a body is directly proportional to the magnitude of the force, and inversely proportional to the mass of the object*. This was a language that was completely foreign to most people of the time, but it soon made a lot of sense, and it told us what would happen to an object in uniform motion if a force were applied to it. We can abbreviate the second law in the formula $a = F/m$, where a is the acceleration resulting from a force (F) on a mass (m).

Newton's third law introduced a new concept called momentum; it is defined as mass × velocity (m × v). And the third law states that *the total momentum of an isolated system of bodies remains constant*. In short, this means that the total momentum before an interaction (for example, a crash) will always be equal to the total momentum after the interaction, assuming there are no outside influences.

It's easy to see that each of these three laws had a tremendous impact on war; they allowed for a better understanding of such things as the recoil of a gun, the impact of a bullet, and so on. But there was still the important question of how and why a bullet or cannon shell returned to Earth. Galileo had shed some light on it, but, for the most part, it was still a mystery. Newton solved this mystery with his law of gravity. It is as follows: *every particle in the universe attracts every other particle of matter with the force that is personal to the product of their masses and inversely proportional to the square of the distance between them.* In mathematical terms this is $F = m_1 m_2 / r^2$, where F is force, m_1 and m_2 are masses, and r is the distance between them.

Newton, of course, was not thinking of this equation in relation to weapons of war in any way, but he was interested in applying it to the moon to correctly predict its period around the earth. When he made the calculation it was close to what was observed, but not exact, and the reason was that the distance to the moon was not known accurately at the time, and neither was the acceleration of gravity.

These laws by themselves would have made Newton one of the greatest scientists of all time, but they aren't the only advances he made. He also made fundamental discoveries in relation to light and optics. In relation to light, for example, he showed that white light was composed of light of all colors, and a beam of white light could be dispersed into a beam of all colors using a prism. He also discovered the laws of reflection and refraction. And he invented the first reflecting telescope (most large telescopes now use reflectors). All his discoveries in light and optics were detailed in his book *Opticks*, which was published in 1730.

In addition, Newton made important contributions to the understanding of sound, heat, tides, and fluid dynamics, and he also made several important discoveries in mathematics in addition to calculus. But perhaps most important, he was the first to formulate and use the scientific or experimental method. In particular, he published the four rules of scientific reasoning. And although Galileo had used the experimental method years earlier, it was Newton who perfected it. He also emphasized the role that theory and experiment played together.

What effect did his discoveries have on war? Some of them had a direct effect, but, for the most part, his laws of motion and gravity had an indirect effect in that they allowed gunners and weapons makers to understand what was going on when a gun was fired, and how the bullet or shell fell to the earth. On the other hand, his optical experiments soon made one of the most critical tools of war possible, namely binoculars. And certainly, his invention of calculus played a very large role.

8

THE IMPACT OF THE INDUSTRIAL REVOLUTION

INTRODUCTION

The Industrial Revolution in England began in 1762 and lasted to 1840. It was one of the most important periods in human history—primarily because of its profound influence on the daily life of average people. In particular, the standard of living was increased, but there were still major problems with conditions.

The era was significant for the military. It changed the way armies were equipped and how they fought, and it introduced mass production, which was something new in the civilized world. Guns, ammunition, and other weapons of war could now be easily produced by the thousands. And of particular importance was that weapons manufacture was standardized so that parts were interchangeable and could easily be replaced.

What role did physics, and science in general, play in this revolution? As it turns out, there is some controversy. There's no doubt that the developments spurred interest in physics; and new branches of physics actually arose as a result. But how much did the earlier breakthroughs of Newton, and the breakthroughs that occurred during the Industrial Revolution, relate to physics? The problem here is the definition of "science," and more particularly "physics." Many argue that "pure physics" made little contribution. And indeed it is true that the major contributions came from applied physics and technology, as most of the advances were actually engineering advances.

Nevertheless, there were dramatic changes in society—mostly for the good, although for the lower class, smog from the new blast furnaces (which used coal as fuel) was something new and unhealthy. And there's no doubt that the Industrial Revolution had a huge effect on war and warfare.

THE FRENCH REVOLUTION

For the most part the Industrial Revolution took place mainly in England, at least in the early years, but looking back in history it's easy to see that its origins were in France. However, it didn't play out fully in France until it was well underway in England.

The origins of the Industrial Revolution can be traced to Louis XIV of France, who ruled from 1643 to 1715. He had the longest reign of any French king—seventy-four years. He became king when he was four years old, but the Queen Mother and her assistant wielded power until he was twenty-one. When he took over, England's navy ruled the seas, and the French army was no match for the highly trained English army. Louis, who was convinced that his power was given to him by God and that he was accountable to no one except God, decided to make France the strongest country in Europe, and to do it he would have to build up its army and its navy. Furthermore, if these were to be first rate, they would have to have first-rate weapons, strategies, and tactics. And he was determined to make this come to pass. Strangely, though, he had no interest in "leading" his armies into war as Adolphus of Sweden had done, and he cared little about new developments in technology, or science in general. His major interest was dancing and partying at his many palaces (he built the immensely plush palace at Versailles). Fortunately, he had a very competent finance minister named John Baptiste Colbert, and he put Colbert to work upgrading the army and navy. And indeed, Colbert did an excellent job; within a few years France had one of the strongest navies and best-equipped armies in Europe. His navy went from 18 outdated ships to 190 ships equipped with all the modern devices known, and his army increased from a few thousand poorly trained men to 400,000 highly trained soldiers, equipped with the best cannons and muskets available at the time.[1]

With all of this available to him, Louis decided he was going to expand the borders of France—in essence, he wanted to conquer Europe and defeat the English in the process. He thought of war as a "sporting event," with himself as commander. He began by attacking Belgium and Holland with his large army. He easily overcame them, but soon other countries saw him as an egotistical aggressor and began to form alliances against him, and, as a result, his losses began to pile up. One of his major losses was the War of Spanish Succession, which started in 1701 and continued to 1714; by the time it ended, France was almost bankrupt. Indeed, throughout much of his long reign he was at war, and by the time he died in 1715 he was highly unpopular.[2]

Even though he was unsuccessful in his expansionist ambitions, he did make the important contribution of starting the Industrial Revolution. It all began with gunpowder. He wanted gunpowder produced fast and efficiently, and the methods that were in use at the time were too slow, so he directed his ministers to build a huge workshop in Paris for producing gunpowder. In it he set up what was probably the first "assembly line" for mass production. Production underwent several steps, with groups of people involved in each step, where each group performed only one operation before passing the product on to the next group. It was a new tactic, and it worked wonderfully. Soon he had warehouses full of gunpowder.

From here he turned to the production of guns—both cannons and muskets—and he set up an assembly line for them. He mass-produced uniforms in another assembly line. From here the new revolution could have spread and made France the greatest industrial nation on earth—but it didn't. By the end of Louis's reign France was nearly bankrupt. As a result, the major part of the Industrial Revolution took place in England.

THE ENGLISH REVOLUTION

The revolution in England, which began about 1760, was fueled mostly by three technical advances: James Watt's steam engine, John Wilkinson's new techniques for iron production, and new techniques in the textile industry. Several developments in the chemical industry, along with the development of new machine tools, were also helpful.[3]

With the advent of steam engines, efficiency increased dramatically. But for the early part of the revolution, industry still relied on waterpower, wind, and horses for driving small engines.

The first successful steam engine came in 1712. It was invented by Thomas Newcomen, based on experiments performed thirty years earlier by Christiaan Huygens and his assistant, Papin. It consisted of a piston and cylinder, with the end of the cylinder above the piston open to the atmosphere. Steam was introduced in the region below the piston. This steam was condensed by a jet of cold water, producing a partial vacuum. The pressure difference between the vacuum and the atmospheric pressure on the other side of the piston caused the piston to move downward in the cylinder. It was attached to a rocking beam that in turn was attached to a water pump.

Newcomen's steam engines were used in England for years for draining the water in mines. But it was the improvements to the design made by James Watt that provided the major breakthrough to the Industrial Revolution. Other things that played an important role in the revolution were the development of new machine tools such as the lathe and various planing and shaping machines. Cylindrical boring machines were also important in relation to war, and they were used for boring cannons. Much of this was made possible, however, by the conversion from wood or charcoal to coal in the large furnaces of the time.

The development of new chemicals, such as sulfuric acid, sodium carbonate, alkali, and so on, was also important. Portland cement was also used for the first time during the Industrial Revolution.

JAMES WATT AND THE STEAM ENGINE

The major breakthrough that made the Industrial Revolution possible was the invention of the steam engine by James Watt. Initially it was just an improvement on Newcomen's model, but it proved later to be much more than that. Born in a well-to-do family in Greenock, Scotland, in 1736, Watt did not attend a regular school during his early years. He was home-schooled by his mother, but later on he did attend Greenock's grammar school. From an early age his skills in mathematics were obvious, but he also liked to build things. When he was eighteen he went to London to study instrument making. Later, he set up a shop in Glasgow as an instrument maker; in particular, he specialized in scales and parts for telescopes, barometers, and various other instruments of the day. His skills came to the attention of the physics and astronomy department at the University of Glasgow, and he was offered the opportunity to set up a small workshop at the university so he could help monitor and repair the instruments used there. As a result he became friends with several of the university personnel; in particular, the well-known physicist Joseph Black (an expert on heat) became his confidant and mentor.[4]

In 1759 another friend, John Robinson, told him about the problems of the Newcomen steam engine, and he was asked to repair a Newcomen engine belonging to the university. Looking into the design of the engine, Watt soon realized that it was extremely inefficient. In short, it was wasting much of the energy it was producing because three-quarters of the heat of the steam was being consumed in heating the engine cylinder during each cycle. The main reason for this waste was the cold water that was injected into the cylinder to condense the steam to reduce its pressure. Much of the energy, therefore, was going into repeatedly heating the cylinder.

Watt redesigned the engine so that the steam condensed in a separate chamber away from the piston. In addition, he maintained the temperature of the cylinder by surrounding it with the steam "jacket." This meant that most of the heat from the steam would now be performing work. This improved the efficiency and power of the engine dramatically. Watt built and demonstrated his new machine in late 1765. Surprisingly, though, even with its obvious advantages and potential, he had trouble finding someone to back him in producing it commercially.

Eventually, though, he was introduced to Mather Bolton, the owner of a foundry near Birmingham, and they became partners. Over the next few years the firm of Bolton and Watt became very successful. Watt continued to improve his machine and soon converted it so that it would produce rotational power. This proved to be a boon in grinding, milling, and weaving. Later he developed a compound engine, in which two or more engines could be used together.

But there was still a problem with the largest engines: the piston in the cylinder did not always fit tightly. This problem was solved by John Wilkinson.

JOHN "IRON MAD" WILKINSON

In 1774 John Wilkinson made a significant breakthrough in the construction of cannons. For years cannons had been constructed of iron that was cast with a core. Any imperfections in the interior were removed by a quick bore job, but this created a serious problem: each cannon was slightly different, and parts therefore had to be custom made. They could not be interchanged from cannon to cannon. Wilkinson showed that casting a solid cylinder and boring a hole in it by rotating the barrel produced much more accurately machined cannons, which would allow for the interchangeability of parts. It also made the cannons much less likely to explode during manufacture. As a result, the production of large cannons was improved. Watt's new steam engines helped Wilkinson produce more large guns with less labor, and Wilkinson's new techniques with iron and steel helped Watt build bigger and better steam engines. The partnership was of great benefit to the English military. Many of the large cannons were installed on ships, helping to make England's navy even stronger.[5]

Although Watt probably didn't realize it, his work was also critical in the development of a new branch of physics, called thermodynamics. Thermodynamics is primarily concerned with studying and improving the efficiency of all kinds of heat engines, and it would soon become an important branch of physics.

The work of Wilkinson and Watt was, without a doubt, critical to the military, but it created a problem. Wilkinson soon began to realize he was indispensable, and he began to think that the British military was not compensating him sufficiently. He was ambitious and wanted to expand his iron and steel empire, but he needed more money, and it looked like his chance of getting much more from the British military was slim. And he also knew that other countries would be eager to pay for his knowledge and technology, France in particular. So, without mentioning it to the British authorities, he met with some French diplomats, and, as expected, they were eager to buy his cannons. But there was, of course, a problem: How could he ship them to France without alerting British custom officials? He got around this by labeling his exports as large iron "pipes." And France paid him so well that he soon became a very rich man.

BENJAMIN ROBINS

While advances were being made in the construction of cannons, advances were also being made in the construction of muskets, particularly in relation to their accuracy. And as it turned out, physics was critical to these advances. Most of these advances were associated with one name: Benjamin Robins.

Robins was born in Bath, England, in 1707 to Quaker parents. His father was a tailor, but the profession brought him little money, and the family was relatively poor. Benjamin's mathematical ability eventually attracted the attention of some of his friends, and a letter was sent to Dr. Henry Pemberton in London. Pemberton sent young Robins a test, and he did so well on it that he was invited to come to London. At the time, Pemberton was preparing a new edition of Newton's *Principia*, and Robins read it along with many other important works in mathematics and physics. By the time he was twenty, Robins was publishing in major journals and was elected a fellow of the Royal Society (a tremendous honor for anyone so young). He continued to publish extensively; in one publication he defended Newton's new "calculus" from several attacks by would-be mathematicians.[6]

A new military academy, the Royal Military Academy, was established in 1741. Robins applied for a position, but surprisingly, despite his impressive qualifications, he was turned down. Some people have said that this annoyed Robins so much that he became determined to show the academy what a mistake it had made, and, as a result, he threw himself into the study of the physics of guns, artillery, and projectiles.

After a careful study of the armaments of the day, he was amazed to find out how inaccurate they were. In some battles as many as 250 rounds of ammunition were fired for every enemy killed. In fact, manufacturers didn't bother to put sights on military muskets because they were never used for individual targets. A volley of bullets fired by a large number of soldiers was the main tactic used at the time.

Robins was determined to find out what the major problems were. As a test he fixed a musket in a rigid clamp and set up targets (paper screens) at distances of fifty, one hundred, and three hundred feet. He then measured how far the bullet had strayed off the target (away from a straight line) for each of the distances. He found that at one hundred feet it was off by fifteen inches, and at three hundred feet it was off by several feet. Furthermore, it was off by different amounts in different directions. So much for accuracy. It was no wonder that it was a waste of time aiming at a target three hundred feet away.

Robins immediately asked himself why the accuracy was so poor. There had to be a logical reason. He finally determined that the problem had to do with the spin of the projectile. Spin was not given purposely; nevertheless, when the bullet emerged from the barrel it had some spin, and this spin was different for each projectile. The reason for it was that the spherical projectile was purposely made slightly smaller than the diameter of the barrel, and as it moved down the barrel it struck the sides of the barrel at various points, and each time it hit, its spin changed. The critical change, however, was the last one just before it emerged from the barrel. This would be the spin it had in flight. And Robins assumed (correctly) that this spin was interacting with the air that the projectile was passing through, and this interaction affected its trajectory.

The next problem, then, was to determine the bullet's speed as it emerged from the barrel, and, if possible, its spin. To determine its speed, Robins invented what is called a ballistic pendulum—one of the most important inventions in the history of gunnery. He began by clamping a musket in position. Directly in front of it he placed a large block of wood that was mounted at the end of a wire or rope so that it could swing like a pendulum. Robins found that when the gun was fired the block absorbed the kinetic energy of the musket ball, and, as a result, the block swung through a few degrees on the supporting wire. The kinetic energy of the bullet was being changed into potential energy in the process. Equating the two types of energy, Robins solve the equation for the velocity of the projectile, and he determined that the musket ball had struck the block with a velocity of 1,139 miles per hour. After all of the years that guns and cannons had been used, this was the first time anyone had any indication of how fast the bullets were going. So it was, indeed, a tremendous feat.

Now that Robins had determined the muzzle velocity of the gun, he had to figure out what happened as the projectile traveled through the air to its target. How was its velocity altered? By moving the block of wood farther away from the barrel, Robins could find out. And it was soon obvious to him that the bullet was losing velocity rapidly. In fact, a bullet lost almost half its initial speed in the first hundred yards. The air through which the musket ball moved was obviously having a serious effect. Scientists and engineers of the time knew that air caused drag on a musket ball, but they didn't realize how dramatic the effect was. The basic problem was the shape of the projectile: a sphere. A sphere was not as aerodynamic as other shapes, and Robins soon realized this. What was the best shape to minimize atmospheric drag? The answer, at least to some degree, was already known. Archers had experimented with different types and shapes of arrowheads over the years, and they had found that the most dynamic arrow tip appeared to be one that was elongated and shaped to a point. But there was a problem with an elongated musket projectile with a point. It would tumble as it moved through space, making things even worse.

Robins analyzed the problem carefully, and, in doing so, he realized that there were actually two problems: finding the most aerodynamic projectile, and getting rid of the random spin as it left the barrel of the gun. Robins soon found that he could solve both of these problems with one significant change. He would give the bullet an elongated form with a pointed end, and he would cause it to spin around an axis through its center in the elongation direction. The best way to do this was to score the inside of the musket barrel with a series of spiraled grooves. In other words, he decided to "rifle" the barrel. But the grooves would only work—give the bullet is spin along its central axis—if the bullet fit into the grooves. In short, the grooves had to bite into the lead as the bullet moved down the barrel. In fact, it was soon shown that the ideal size for a bullet was slightly larger than the diameter of the barrel.

Finally, Robins drew up plans for a breach-loaded gun; this was a gun in which the breach could be opened so that the charge of powder and a bullet could be inserted into the barrel. With the powder and bullet in place it could then be securely closed so that it was ready for firing. This was a significant breakthrough, but it wasn't actually used until several years after Robins's death. Rifled muskets were a problem technologically, so they didn't really take off for another few years. Nevertheless, Robin's breakthrough revolutionized military warfare, and it soon made England one of the strongest countries in Europe.

THE FLINTLOCK

The basic musket of the time also underwent a significant change during this era. The change was introduced early in the seventeenth century, and by 1660 the new musket became the major European military musket, and it continued to be used until about 1840. Most muskets of the time had smooth barrels and fired lead balls. They had a range of about 150 yards, and they weighed about ten pounds. When rifling began to be used on the barrels it gave them considerably better accuracy and a much greater range. But for the most part, rifled muskets were used only by sharpshooters (or what we might call snipers). The problem was that the rifled musket took much longer to load because of the use of tight-fitting bullets. Furthermore, the barrels of the rifled muskets (now called "rifles") had to be cleaned after each shot, and this took too much time.[7]

The major difference in this new rifle, called the flintlock, compared to the wheel lock, was the use of a piece of flint to create sparks. A person firing a flintlock would cock the gun by using his or her thumb to pull a lever back against a strong spring (see diagram). At the end of the lever was a piece of flint with a point on it. As in the case of previous guns, the flintlock had a flash pan that was loaded with finely ground gunpowder. When the flash pan was primed it was closed. On the top of the flash pan was a steel striking plate called a frizzen, and when the trigger was pulled, the spring would force the flint down toward the plate. When it struck the frizzen it caused the pan to open. And as it slid down the frizzen it created a shower of sparks. The sparks would ignite the powder in the flash pan. As a result, a flame would flare down through the touchhole and ignite the charge in the barrel.

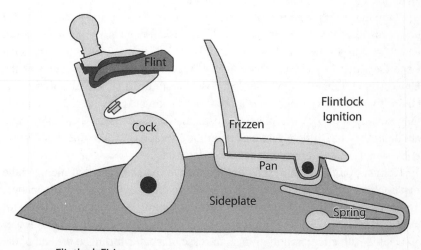

Close-up of the flintlock mechanism.

The flintlock mechanism was used on both rifles and pistols. By now, in fact, military pistols were becoming quite common; they had a relatively short range, but they were easy to handle, which made them popular with the cavalry. The smaller flintlock pistols were about six inches long; larger ones were about sixteen inches long. One of the most popular pistols was the Queen Anne pistol. This elegant and beautifully designed gun was frequently the weapon of choice for dueling, which was a popular method of solving arguments at the time. Indeed, some of the pistols even had two, three, or more barrels for quicker shooting.

Although they were a significant improvement over matchlocks and wheel locks, flintlocks had problems of their own. The flint had to be sharp or the gun would misfire. Also, the flintlock was vulnerable to moisture and accidental firing. In addition, occasionally the gun would explode in soldiers' hands.

CHRISTIAAN HUYGENS

Getting back to the physics of the era, we have one of the greatest physicists to appear after Newton. Unlike Robins and some of the scientists we've discussed so far, Huygens was born into a highly respected and relatively rich Dutch family; his father was a diplomat and part-time natural philosopher who played an important role in Christiaan's early education.[8]

Christiaan was schooled at home by some of the best local teachers until he was sixteen. It was obvious that he had considerable ability in mathematics, and this came to the attention of a family friend, the eminent mathematician Rene Descartes. Descartes encouraged Huygens to study mathematics at the University of Leiden. And indeed, Christiaan studied mathematics as well as law at the university, beginning in 1645. Over the next few decades he made numerous discoveries in physics, mathematics, and even astronomy, and although they had little effect on the military weapons of his time, his discoveries would have a tremendous influence later on.[9]

In addition to fundamental contributions to mathematics that included the first book on probability theory and solutions to many of the basic mathematical problems of the day, Huygens made major contributions to physics. In 1659, for example, he derived a formula for the force (now called the centripetal force) associated with circular motion, as in the case of a ball whirling at the end of the string. His formula stated that $F = mv^2/r$, where m is mass, v is velocity, and r is radius. He also made fundamental discoveries relating to the elastic collision of two bodies; he was the first to show experimentally that the total momentum before the collision is always equal to the total momentum after the collision (as predicted by Newton's third law). In addition, he invented the first pendulum clock, and he was the first to derive the formula for the period of a pendulum. He also devised new methods for grinding lenses and making telescopes, and he used telescopes to discover the largest moon of Saturn, now named Titan. He also observed Saturn's ring and predicted correctly that it was thin and not attached to the planet.

In the area of physics, however, he is probably best known for his wave theory of light, which he proposed in 1678. A few years later Newton suggested that light was composed of tiny particles he called corpuscles, and for many years there were two theories about the nature of light: one based on waves, and one based on particles. In 1801 Thomas Young showed that Huygens was correct, but today's quantum physics has embraced the concept of wave–particle duality, since light seems to exhibit both properties, depending on one's method of observation.

Huygens also developed a balance spring for clocks and watches that is still used in some modern devices, and in 1675 he patented the first pocket watch. In 1673 he began experimenting with a combustion engine, which he fueled with gunpowder. It was not successful, but he designed a simple form of steam engine that was helpful to James Watt in his work.

PHYSICS AND THE INDUSTRIAL REVOLUTION

As mentioned earlier, a number of scholars have suggested that pure science (including physics) played only a minor role in the development of the Industrial Revolution. But if you look at the overall picture, it's easy to see that many fundamental discoveries in physics took place during this time, most notably those of Huygens. In addition, both the Royal Society in England and the French Academy in France were formed. The goal of both organizations was the enhancement of pure and applied physics. And although enhancement of the military was not a major goal of either, it was no doubt a secondary goal.

Another important advance during this time, which no doubt helped Watt's steam engine, was a formulation of the fundamental gas law now known as Boyle's law. It states that the product of pressure and volume is constant for a given mass of confined gas, as long as the temperature is constant. This means, for example, that when the volume of the gas is halved, the pressure is doubled, or if the volume is doubled, the pressure is halved. It was first stated by Robert Boyle of England in 1662.

Watt's work, in turn, was critical to military development in that it made the production of both cannons and muskets much more efficient. Furthermore, as mentioned earlier, a major branch of physics, namely thermodynamics, was created as a result of Watt's work on the efficiency of engines.

Huygens's work on collisions and the centripetal force was also helpful in relation to the development of weapons of war, but it was his work on light, and his suggestion that light was a wave, which would eventually have tremendous implications. A few years later, with the work of Maxwell and Hertz, it would lead to the discovery of the electromagnetic spectrum, which would eventually have a very large impact on war.

And finally, Benjamin Robins's advances in gunnery were based primarily on physics, and they would dramatically change war over the next few decades.

9

NAPOLEON'S WEAPONS AND
NEW BREAKTHROUGHS IN PHYSICS

While the Industrial Revolution was sweeping across Europe and changing the structure of society, much of Europe was at war. In France the French Revolution began in 1789 with the storming of the Bastille in Paris; three years later Louis XVI was beheaded and the revolution was in full swing with twenty to forty thousand people beheaded in two years. France was bankrupt, but surprisingly, several important scientific advances were being made, and they were mainly associated with one man, Antoine Lavoisier. Then, to the amazement and grief of many, Lavoisier was beheaded, and scientific advances stopped. During the Napoleonic Wars, few scientific advances were made, and most efforts in this direction were focused on the military.

THE FRENCH REVOLUTION

When Louis XVI came to the throne in 1774 he was surprised to find that the military had a serious problem: lack of gunpowder. The warehouses were almost depleted and sources were drying up. In particular, the major component of gunpowder, saltpeter, was in short supply. For years it had come mainly from stables and old buildings. Louis decided that action was needed; he appointed Antoine Lavoisier, the best-known chemist in France, to head up a commission. As it turned out, Lavoisier was the right man for the job. He went to work immediately, offering a cash reward for any new advance in the production of saltpeter or gunpowder in general. He quickly got rid of all outdated procedures and looked carefully at the formula for gunpowder to see how it could be improved. And within a short time he had made significant changes. Four years later the formula for gunpowder had been improved so much that foreign nations were eager to buy it from France, and the formerly empty warehouses were now full.[1]

But the improvement of gunpowder was only one of Lavoisier's achievements. He made so many discoveries in chemistry that he came to be known as the father of modern chemistry. And although he was primarily a chemist, his breakthroughs had a strong influence on physics. In particular, he made important contributions to physical chemistry and thermodynamics. He burned sulfur and phosphorus in air and proved that the product of the combination weighed more than the original elements. He then went on to show that the weight gained was lost from the air, and that, as a result, mass was conserved, leading to the law of conservation of mass.

He also showed that air has two components, one of which causes metals to rust. He named this component oxygen. Then he showed that the component of air discovered years earlier by Henry Cavendish was present in water. In particular, oxygen and the new component, which he called hydrogen, appeared to make up water. Furthermore, he noticed that nitrogen was also a major component of air.[2]

For years the accepted theory of combustion was the "phlogiston theory," in which it was assumed that a mysterious element called phlogiston was contained in all combustible bodies and was released during combustion. Lavoisier showed that the theory was wrong. It was oxygen that was the key concept in combustion. He was also the first to make a comprehensive list of the known elements (or substances that could not be broken down), namely oxygen, nitrogen, hydrogen, phosphorus, mercury, zinc, and sulfur.

It was perhaps not a surprise that Lavoisier was arrested in 1794. Many well-known scientists of the time pleaded with the assembly to spare his life, pointing out his numerous contributions to the science of France. But the judge, according to an apocryphal story, replied, "The Republic needs neither scientists nor chemists; the cause of justice cannot be delayed."[3] He was beheaded on May 8 at the age of fifty. The well-known mathematician Lagrange stated, "It took them only an instant to cut off that head, but France may not produce another like it in a century."[4]

It wasn't long, however, before elected officials realized that they had gone too far. They released Lavoisier's widow from jail and returned his belongings to her, and years later a statue was erected in Paris in his honor.

JEAN-BAPTISTE VAQUETTE DE GRIBEAUVAL

The French army sank to new lows during the revolution, and even in the years before. Its artillery had grown outdated compared to that of other nations, and the army was ill-trained and generally in disrepair after years of neglect. Jean-

Baptiste Vaquette de Gribeauval, a young lieutenant colonel, was loaned out to the Austrian army (an ally) upon the outbreak of the Seven Years' War, and he was amazed to discover that French cannons and other guns were much inferior to Austrian guns. Within a few years, however, Gribeauval had changed the French cannon into a lighter, more powerful gun with an equivalent range.

Gribeauval did for France what Wilkinson had done for England. At the time, cannon barrels were cast by pouring molten iron or bronze around a clay cylinder. When the material cooled, the clay was removed and the interior was polished. This left guns that differed in size, which made it difficult to produce tight-fitting cannonballs. Because of this, much of the firepower of the explosion was lost. Gribeauval instituted a new system that involved casting cannonballs from a solid block then drilling a hole through the block. As a result, cannons were made more precisely so that parts from one would fit another. Furthermore, cannonballs now fit the barrel much better. This allowed manufacturers to make much lighter cannons that were easier to maneuver but had the same range, or better, than the old cannons. Furthermore, Gribeauval trained his officers to use the cannons in the best way. One of these officers was Napoleon Bonaparte.

Napoleonic cannon.

NAPOLEON AND HIS WEAPONS

Napoleon studied physics along with mathematics and astronomy in military school, and he knew the importance of science to war. And he made some effort to make sure France was at the forefront of technology; he made the École Polytechnique into a military school, and it eventually became one of the most technologically advanced schools in Europe. And he made sure the weapons he was using—particularly the cannons—were the best in Europe. Most of his victories came about, however, not because of his new innovations in weapons, but because he used new and clever strategies and tactics. He was not responsible for any "miracle" weapons that used new breakthroughs in physics, and there's no indication he took a lot of interest in physics or science in general. If it didn't help the war effort he had little interest in it. And in some cases he made mistakes when new innovations were presented to him. One example was the balloon, which was developed in 1782. One of his scientific advisers told him about balloons in 1800 and pointed out that they could be used for surveillance of the enemy and possibly even for dropping bombs on them. Napoleon was curious about the idea at first, but he soon lost interest, and nothing came of it. In addition, even though "rifling" of musket barrels was known at the time, and rifled muskets were known to be more accurate and have a range over three times that of smoothbore muskets, Napoleon didn't like them. They were too slow for his taste, and he stuck with the smoothbore guns for the most part.[5]

The main weapons of the time were muzzled-loaded, smoothbore muskets, along with pistols, bayonets, swords, and pikes. And Napoleon, strangely, was particularly attached to the bayonet, mostly because it was a very effective terror weapon. There's no doubt, however, that his most effective weapon was Gribeauval's new cannons. It was the cannons, along with his new tactics and strategies, which produced victory after victory during his early years as a commander. More than anything, Napoleon looked for his enemies' greatest vulnerabilities and quickly took advantage of them. The speed and efficiency with which he maneuvered his men and cannons on the battlefield surprised and frequently demoralized his enemies. One of his major tactics was to "feint" from the front while he was secretly surrounding his enemies. Then, to their surprise, he would attack from the rear and sides, cutting off their communications and supplies.[6]

Napoleon rose rapidly through the ranks as a young officer, and in 1796, when France attacked Austria, he was given command of the French army in Italy, and in battle after battle he was victorious, returning to France in 1797 as the nation's hero. And it wasn't long before he was on the march again; in May 1798 he left for a campaign in Egypt, where he hoped to draw the British into battle by threatening their commercial interests in the region. He had no problems with the ill-equipped Egyptian army and easily slaughtered over two thousand Egyptian troops with light losses on his side.

But his problems with the British were only beginning. Admiral Horatio Nelson's battleships swept into the bay that contained Napoleon's ships and sank most of them, leaving the French army with no escape. Napoleon quickly abandoned his army and managed to get back to France with a few bodyguards and army marshals. And surprisingly he was greeted as a hero for his victory in Egypt, which contributed to his election as first consul in 1800. Napoleon was now leader of the best-equipped army in Europe, and he soon took advantage of it. He attacked the Austrians first, and with his emerging new tactics and skills he quickly defeated them at Ulm in October 1805. Then most of the Austrian army surrendered to him. Next came the Russians and the remains of the Austrian army at Austerlitz. Napoleon quickly cut the army in half and surrounded them, in the process inflicting tremendous casualties.

Napoleon Bonaparte.

The following year he attacked Prussia, one of the strongest, best-equipped armies in Europe, and to everyone's surprise he defeated it. He now had much of Europe under his command, but he had not yet overcome his major foe— the British. And they were still a thorn in his side. The major problem was the British navy; the French navy was no match for it, and unless he invaded the mainland of England, his armies would do him no good.

After a couple years of relative peace, Napoleon decided to invade Spain, which was known to have a relatively weak army, and, indeed, he soon overcame it. But suddenly he was faced with a new type of war: guerrillas hid in the mountains and ambushed and sabotaged his army again and again. In addition, the British soon came to the assistance of the Spanish, and to make things even worse, Austria was threatening France. Napoleon therefore withdrew, leaving his troops again to a slow but decisive disaster. Over the next few years many of his best troops were killed.

But his biggest defeat came in 1812 when he attacked Russia. With one of the largest armies ever assembled, numbering approximately six hundred thousand troops, Napoleon expected an easy victory. As it turned out, his army was too big, and the Russians were too cunning. They retreated as Napoleon swept across the land, getting closer and closer to Moscow. And they knew what they were doing; they were waiting for the harsh and cruel Siberian winter. Furthermore, as the Russians retreated they applied a scorched-earth tactic to the areas they passed through, and, as a result, the French, who depended on foraging for food in conquered lands, found themselves short of everything they needed for survival. Napoleon was sure, however, that once he got to Moscow there would be enough food, and the war would soon be over. But the ever-cunning Russian army retreated past Moscow, leaving nothing in it. Napoleon entered the city and waited for a surrender from the Russian general, but it never came. After a month of near starvation, he and his troops left Moscow for the long trek back to France in the dead of winter. The war had begun with over six hundred thousand troops under Napoleon's command, and by the time they got back to France there were fewer than thirty thousand left. Napoleon's days of glory were over, but he was still in power.[7]

Back home Napoleon managed to assemble a new army of 350,000. But now his foes were unified; Russia, Prussia, Austria, Great Britain, and Spain had formed a new coalition. Nevertheless, Napoleon did have a few victories, but at the battle of Leipzig his army was reduced to seventy thousand. Paris was soon surrounded, and it was captured in March 1814. The victors exiled Napoleon to the island of Elba in the Mediterranean. Amazingly, he escaped from Elba in February 1815 and returned to the mainland, where he was hailed as a hero, and for another hundred days he again governed France. Then he met the British general Wellington at the Battle of Waterloo and suffered his greatest defeat. This time he was banished to St. Helens in the Atlantic Ocean. He died in 1821.

COUNT RUMFORD

Although few discoveries in physics were made in France during the Napoleonic era, important advances were being made elsewhere, and most of them were related to heat and thermodynamics. Count Rumford (his honorary title) was born Benjamin Thompson in Woburn, Massachusetts, in 1753. Early on he worked for the British army, conducting important experiments on gunpowder. His results were published by the Royal Society of England in 1781. He also began a series of experiments on heat about the same time.[8]

When the American Revolutionary War was over he left for London. After four years, however, he moved to Bavaria, where he continued his experiments on heat and light. He was recognized by the Bavarian government in 1791 and made a count of the Holy Roman Empire, with the title Count Rumford. The name he chose was the name of the town in New Hampshire where he had been married.

His major scientific interest for many years was heat. Early on he devised a method for measuring the specific heat of solid substances. Specific heat is the amount of heat needed to raise a given quantity (e.g., 1 gram) of a substance by one degree. He delayed publishing his results, however, and was disappointed when someone else published the same result before he did.

His most important discovery, however, took place while he was in Munich. He was put in charge of manufacturing brass cannons, and as he watched how they were being made he was amazed by the amount of heat that was produced by the boring. Water was used for cooling, but it boiled rapidly. He decided to measure the amount of heat produced in the process. Setting up a specially shaped cannon barrel that could be insulated against heat losses, he immersed the drill and barrel in a tank of water and measured the temperature increase of the water as the boring took place. This allowed him to determine how much heat was being produced. But he went a step further: he calculated how much heat was produced for a given amount of mechanical work. We now refer to this as a mechanical equivalent of heat. His value is somewhat higher than the value we now accept (4.18 Joules per calorie), but it was an important first step, and it established an important relationship in physics.

In performing the experiment Rumford also showed that no physical change had taken place in the material of the cannon, and the supply of frictional heat seemed to be inexhaustible as long as the boring continued.

Rumford also made important contributions to photometry, the measurement of light. In particular, he introduced the light unit, the standard candle.

NEW BREAKTHROUGHS IN PHYSICS

While wars were raging, other important discoveries were being made in physics, most notably in the study of electricity and magnetism. It would take years to understand the new phenomena thoroughly, and to apply them to useful devices, but there's no doubt that they eventually had a huge effect on warfare and weapons, and on everyday life.

In the early 1730s the French physicist Charles du Fay discovered that electrified objects are sometimes attracted to one another and sometimes repel one another. He postulated that there are two types of electrical fluid, which he called vitreous electricity and resinous electricity (later called positive and negative electricity). He also noted that some materials conduct electricity better than others; he referred to this property as "contact electrification."

A few years later, in 1746, the statesmen and scientist Benjamin Franklin became interested in electricity and carried out experiments using a Leyden jar (a jar with a brass rod down its center that could be used to store electrical charge). He wondered if the lightning bolts during thunderstorms were related to the sparks that could be produced near the ball at the top of the Leyden jar. To satisfy his curiosity he flew a kite during a thunderstorm, equipping it with a pointed wire connected to a silk thread, and tying a metal key to the other end of the thread. As expected, when he put his hand near the key, it sparked in the same way that Leyden jars did. Franklin was now sure that the "electric fluid" of the Leyden jar was also present in the storm clouds.

The French physicist Charles Coulomb began looking into the problem of the attraction and repulsion of electrified bodies in the early 1780s. He was particularly interested in the force between them. If they attracted or repulsed one another there had to be a force associated with the phenomena. He constructed a very sensitive device called a torsion balance that allowed him to measure the magnitude of the force, and he found that it was proportional to the inverse square of the distance between the two charges, and proportional to the product of the two charges. We now write this as $F = q_1 q_2 / r^2$, where q_1 and q_2 are the magnitudes of the two charges and r is the distance between them.

The scene then switched to Italy, where the physician and physicist Luigi Galvani began to take an interest in the new field of "medical electricity" in about 1790. One day Galvani was skinning a frog on which he had been experimenting with static electricity. His assistant had touched a metal scalpel to a nerve on the frog's leg with a charged Leyden jar nearby. When the scalpel touched the nerve, the dead frog's leg jumped, as if alive. This observation surprised him, and he published it in 1791. He assumed that the jerking was caused by an electrical fluid in the nerves, and he called the phenomena "animal electricity."[9]

Soon after the result was published, another physicist in Italy, Alessandro Volta, read about it and repeated the experiment. He noticed almost immediately that a frog was not needed; the only thing needed was two dissimilar metals and a moist conductor (they replaced the frog leg). And within a short time he went a step further, showing that a series of several bimetallic strips and moist conductors worked even better. Volta continued working on his new device, which he called a pile, using disks of silver and zinc on top of one another with cardboard disks soaked in salt water between them. With the new device, which we now know as a battery, the continuous flow of electrical current was created for the first time.

Once scientists had a "current" of electricity flowing in a wire conductor, many physicists began experimenting with it. Among them was the German physicist Georg Ohm. Ohm soon found that the current that flowed along a wire between two points depended on the "resistance" of that section of wire. His law is now referred to as Ohm's law. Current is now measured in terms of a unit called an ampere, and resistance is measured in units called ohms. Mathematically, his law can be stated as $V = IR$, where V is the voltage between the two points, I is the current, and R is the resistance.[10]

But there was still a serious problem. Because electricity had so many properties that were similar to magnetism, it appeared that they had to be related, but no one could prove it. In 1813 the Danish physicist Hans Christian Oersted became interested in the problem, but after experimenting for several years he was unable to find a connection between the two. Then one day in 1820 he was giving a lecture; during the lecture he was turning an electrical current off and on. Nearby

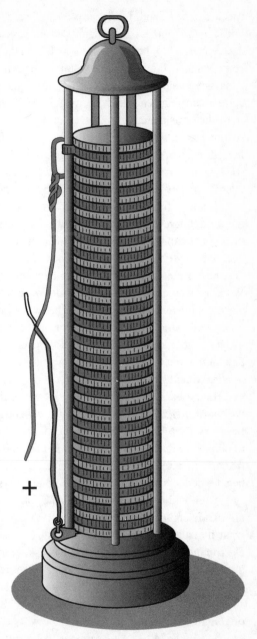

The voltaic pile of Volta.

was a compass, and, to his surprise, he noticed that his actions were having an effect on the compass. He brought the compass up to the wire, holding the compass needle parallel to the wire. When he turned the current on, the needle moved to a perpendicular direction. As a result, he determined that an electrical current has a magnetic field associated with it. The magnetic field surrounded, or circled, the current-carrying wire, and the magnitude of the field weakened as the distance from the wire increased.[11] Oersted published his results in July 1820, and they soon caused a sensation. There was now proof: electricity and magnetism were, indeed, related. In particular, an electric field would produce a magnetic field. It was also soon found that a moving magnet could produce electrical current. The interaction between the two fields is now referred to as electromagnetism.

Within a few weeks of the publication of Oersted's discovery, the French physicist Andre Ampere read about it. He verified Oersted's work and went on to conduct experiments on the fields around the wire. He showed that two parallel wires carrying electrical currents attracted or repelled each other, depending on whether the current flow was in the same direction or in opposite directions. Working out the details of the interaction, he showed that the forces between them obeyed an inverse square law. He then went on to develop the "right-hand rule" for current, which says that if you grasp a current-carrying wire with your thumb in the direction of the current and close your fingers, your fingers will point in the direction of the magnetic field. He was also the first to develop a solenoid—a coil of wire wound in a spiral that created a magnetic field down its center.

But perhaps the greatest shining light of the era was Michael Faraday, who was born in 1791. He was basically self-educated; at fourteen he was apprenticed to a local bookbinder, which brought him into contact with large numbers of books. He read as many of them as he could in his spare time, and he was particularly inspired by one that described the new phenomena of electricity. Later he attended lectures given by the eminent physicist Humphrey Davy.[12]

After reading about Volta's pile he constructed one for himself, and in 1821, after Oersted announced his discovery, Faraday built two devices that produce what he called "electromagnetic rotation." They were simple versions of the electric motor. Then in 1830 he began asking himself if a magnetic field that was already in existence could produce an electrical current. To find out he wound a wire around an iron ring and attached the two ends to a battery, producing a solenoid, then he placed a switch in the circuit so that it could be turned on and off. On the opposite side of the ring he wound several loops of another wire and attached its ends to an instrument that measured current, called a galvanometer. Faraday then turned the switch off and on several times, expecting to see a current in the second wire. To his disappointment, however, there was only a tiny current that lasted for a fraction of a second. Experimenting further, he finally determined that it was not the existence of the magnetic lines that created the current but rather the motion of the magnetic field across the wire. He soon showed that if he merely pushed a magnet into a coil of wire, it would produce a current in the wire. We now refer to this as electromagnetic induction.

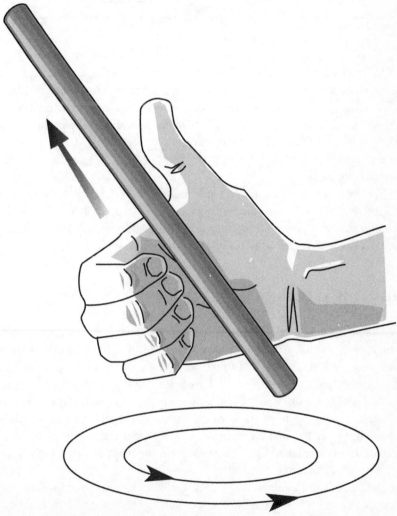

The right-handed rule for the direction of the magnetic field.

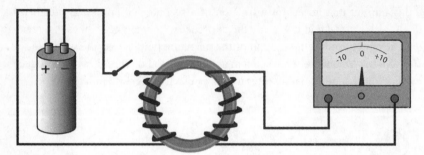

Faraday's induction coil.

In 1845 Faraday also discovered that certain materials exhibited a weak repulsion from a magnetic field. He called the phenomena diamagnetism. In addition, he showed that magnetism could affect a ray of light, demonstrating an apparent relationship between magnetism and light. Finally, in his later years, he proposed that the electromagnetic force actually extended into the empty space around a conductor in the form of "lines of flux." We now refer to them as electric field lines.

Within a few years Faraday's work led to two very important inventions: the electrical generator and the transformer. The electrical generator could be used as a source of power for industry, and the transformer was soon being used extensively for changing or adjusting the voltage of a source.

HOW THIS AFFECTED WARFARE

Even though the above discoveries were some of the most important in the history of the world, it was several years before they were used in war. But when they were finally applied to the technology of the time, they created a revolution. Electrical generators and electrical motors were soon developed as a result of these discoveries, and they eventually played a large role in the development of weapons of war. Large electrical generators eventually replaced steam engines as the major source of power. Power plants sprang up across much of the civilized world, spurring production on many fronts. Weapons were soon produced at a tremendous rate.

But one of the major outcomes of the breakthroughs in electricity and magnetism was a sudden new interest in science and technology, with physics at the forefront. Most countries began to realize how important physics and other sciences were to warfare and the development of new weapons. Pure science had been frowned upon prior to this, but government officials increasingly began to realize the importance of pure science, and physics in particular, in relation to the technology of military applications.

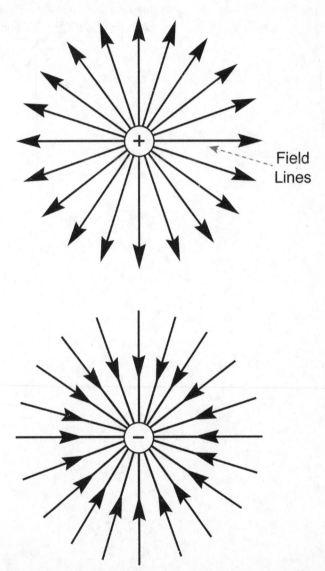

Faraday's lines of flux, or electrical field lines for a positive charge and a negative charge.

Many new universities sprang up in England, France, and Germany, and also in America. And a new emphasis on science and math, including physics ensued. Other countries, such as Japan and Russia, also soon came to the same conclusion.

The first "dynamo"—an electrical generator capable of producing power for industry—was built in 1832. Then came the electrical telegraph and fuel cells.

THE AMERICAN CIVIL WAR

Historians refer to the American Civil War as the first truly modern war because it used so many new and greatly improved methods and weapons. Many important developments in physics and weaponry occurred just before and during the war. Although weapons had been produced in large numbers in Europe earlier, the first true mass production of some of the most deadly weapons that had ever been used in war occurred during these years. In addition, developments in physics and other sciences led to the wartime use of the electric telegraph, electric generators, surveillance balloons, better and larger ships, torpedoes, and significantly improved telescopes.

DEVELOPMENT OF THE PERCUSSION CAP

The flintlock was still being used by many at the beginning of the war. But a new invention that would supersede it had already been made. In 1800 Edward Howard, an English chemist, had discovered a highly explosive material called mercury fulminate. He hoped it might replace gunpowder, but to his disappointment, when he tried it in a musket, it blew the barrel apart. It was too explosive.[1]

Reverend Alexander John Forsyth of Scotland followed up on this discovery in 1807. Like Howard, he thought there was a definite need for a new gun explosive. But he was more concerned with the mechanism used to set off the charge, namely the spring-loaded flint. A spark was needed to trigger the primer, and things didn't work well during damp weather or rainy days. He decided to try mercury fulminate as a primer. It had a serious advantage: it didn't take a match to make it explode. In fact, it could easily be made to explode just by banging it with a small hammer. So he developed a spring-loaded device that would strike some mercury fulminate in a small paper cartridge. The cartridge was attached to a tube that led into the gun barrel. The flames from the explosion went down the tube and ignited the powder that propelled the bullet. He was pleased at how well it worked.

Forsyth and others continued to work on the new system over the next few years. First a paper cartridge was used with mercury fulminate sealed between paper sheets. By 1814, small iron caps were used to contain the fulminate, and later this was changed to copper. Eventually a copper or brass cartridge was used that contained both the bullet and the gunpowder.

The new percussion system had many advantages: weather was no longer a problem, and it was much faster and more convenient to load. We can, in fact, say that the percussion cap revolutionized muskets and pistols. Armories throughout the world soon began converting flintlocks to fulminate-fired weapons. And indeed this also happened in America, but not until the Civil War was well underway.[2]

A significant breakthrough had been made, but there were still problems. And one of the main ones was related to reloading the gun. Most muskets were single-barreled and had to be reloaded after each bullet was shot. A number of inventors tried to use side-by-side barrels, each with a bullet in it, but this didn't work well for warfare. Something better was needed. And one of the first to try to do something about it was Samuel Colt.

Colt's main concern early on was the pistol. How could it be made to shoot several bullets one after the other? He began by making a wooden model. His idea was to have a revolving cylinder with several chambers. Bullets could be placed in the chambers, with one of the bullets lining up with the barrel, and when the trigger was pulled the cylinder would revolve and another chamber, with another bullet ready to fire, would line up with the barrel. What he had was several bullets in a cylinder that rotated; his most famous gun had six bullets.[3]

Colt took his new invention to the army in the early 1840s, well before the Civil War, but the army had little interest in it. Despite the setback, he managed to scrape together enough money to set up a factory at Patterson, New Jersey. But at this stage his "six shooter" was still crude, and it wasn't too successful. He continued working on it, however, and gradually reduced the number of parts in it down to seven.

Finally, the gun began attracting the attention of others—particularly the Texas Rangers. Colt's invention was an ideal weapon for someone on a horse. In 1847 Colt set up another plant for his new, improved handgun at Hartford, Connecticut. It was a .31-caliber weapon that was relatively light compared to other pistols. Soon the new revolvers were being mass-produced. Using systems developed during the Industrial Revolution, he standardized all the parts within the gun so that the parts from one gun would fit all guns, and over the years he produced 325,000 of them.

But the army, even later in the Civil War, was slow to take up his gun.

THE MINIÉ BALL

The percussion cap was a dramatic step forward, but within a short time after its development an even greater invention made the musket into a much more deadly weapon with greater range and accuracy, and it doomed the smoothbore musket forever.

The new development began in 1823 in India when a British officer, Captain John Norton, noticed something strange. The Indian natives used a tube for pro- jecting darts at their enemies, and when they got ready to fire, they began by blowing into the barrel. He discovered that they were doing this to create a foam that would fill the barrel and effectively seal it, so that when the dart was shot, the force on it would be much greater.

In 1836 a London gunsmith improved on Norton's idea by inserting a wooden plug in the base of the bullet so it would expand when shot. This helped, but the real advance came when a French army captain, Claude Minié, improved the design using a hollow cylindrical base. The bullet was now cone-shaped, similar to our modern bullets. So, even though it was called a Minié ball, it was not shaped like a ball. At first the Minié ball had a round cup in the base, and when the powder exploded the cup forced the lead outward to fill the barrel. What was particularly important about this was that the bullet was now fitting snugly into any rifled grooves that were in the barrel.[4]

Spiraling rifled grooves had been used for years, but for a snug fit, which was required, the bullet had to be slightly larger than the interior of the barrel, and it had to be pounded down to a position just above the powder, and this was a slow process. The Minié ball, on the other hand, could just be dropped into the barrel, and this was much faster. And as the Minié bullet caught the grooves as it exited it was forced into a spin, and as a result, it left the barrel with a very high spin rate.

To see why a spinning bullet was so revolutionary, we have to look at the physics of a spinning object. When an object of any type rotates, it rotates around an axis, and this axis of rotation acquires a special status. In the case of a bullet in flight (shot from a rifle) there are two motions we have to consider: its translational motion (that gives it its trajectory) and its rotational motion. It has both at the same time, in the same way a curving baseball does. A pitcher purposely gives a baseball a spin to curve its path so that it is more difficult for a batter to hit.

How do we deal with a spinning object? First of all, it's easy to see that it spins about an imaginary line called its rotational axis, and we refer to its spin rate as its angular speed (or angular velocity, for a particular direction). Speed of rotation is usually measured as so many revolutions per minute (rpm). Scientists also use another unit, which is particularly convenient in physics. To define it we first have to define what is called the radian; it is $360°/2\pi$, which is approximately $57°$. The unit, radians per second, is commonly used in physics.

So what does it take to set an object in rotational motion—in other words, to make it spin? It obviously takes a force. This takes us back to the concept of inertia. Remember that according to Newton's first law, an object in motion remains in uniform motion with a constant speed in a straight line unless acted upon by a force. In short, a body in motion has inertia, and it takes a force to overcome this inertia. Inertia is therefore a kind of "unwillingness" to change. In the same way, a spinning body has rotational inertia, and it prefers to maintain this inertia. In effect, it takes a force to change it. In the case above, however, we are dealing with a rotational motion, so the force is a rotational force, and we refer this force as torque. (You apply torque every time you turn a doorknob or open a jar.)

If we look at a spinning disk, however, it's easy to see that the "linear speed" (e.g., feet per second) across the disk varies. The speed at a point near the edge is obviously greater than the speed at a point near the center. This means that for a spinning object the speed at various points throughout the object increases as the distance from the spin axis increases. Because of this, ordinary (or linear) force f, and rotational force, or torque, which we denote by τ, are related. This can be expressed as $\tau = f \times r$.

Getting back to rotational inertia, it's easy to show that a spinning object prefers to maintain a spin in a particular direction. Assume you have a bicycle wheel with a handle on its axis so that you can hold on to it with your hands. If you set the wheel spinning, then try to twist it, you will find that it's very difficult to turn. In short, the wheel wants to keep spinning in the same direction. This means that a bullet spinning around an axis along its elongated shape, and traveling in a certain direction, prefers to maintain this direction. Spin therefore "stabilizes" a bullet in flight. As it turns out, it also decreases the effect the air around it has on it (i.e., air resistance). Because of this, the Minié ball was much more accurate and had a greater range.

It's important to note that applying torque to a nonrotating object gives it an angular acceleration, where units of angular acceleration are radians/sec^2. And again, the relationship between linear and angular acceleration is given by the formula $\alpha = a/r$, where α is angular acceleration and a is linear acceleration. Finally, in the same way that we have linear momentum, we also have angular momentum, and the conservation principle: *the total angular momentum of an isolated system remains constant.*

With a rifle that has four to eight spiral turns down the interior of its barrel, a Minié bullet will exit with a spin of up to twenty thousand revolutions per second, which gives it tremendous stability compared to the nonspinning spherical ball used in muskets.

A REVOLUTION IN RIFLES AND CANNONS

As discussed earlier, the smoothbore musket was ineffective at ranges greater than one hundred yards, and even at one hundred yards it wasn't very accurate. Lines of infantrymen with smoothbore guns could approach one another to within about one hundred yards and not have to worry about being hit. And even after rifling was first used, few rifles were used in war. They were just too slow to load, and their barrels became clogged with powder residue quite easily. Rifled muskets actually existed well before the Civil War, but they were used primarily by frontiersman such as Daniel Boone, where speed of loading was not critical. In 1750, however, the famous Kentucky rifle was produced, and it was used by many well-known frontiersmen. It was quite accurate; frontiersman could easily hit a turkey at two hundred yards. Furthermore, the Kentucky rifle was used by Americans in the Revolutionary War; indeed, it spooked the British cavalry because an American sharpshooter with the Kentucky rifle could shoot the horse out from under cavalrymen at four hundred yards. The British, in fact, had to quickly build a gun that would match it; and they did—it was called the Ferguson rifle.

With the new developments—in particular, the Minié ball—came several new and much deadlier rifles. At first the rifles were still muzzled-loaded; nevertheless, they were much deadlier than the smoothbore muskets. The major rifles of the Union were the .58 caliber Springfield musket and a .69 caliber rifle developed at Harpers Ferry called the Harpers Ferry rifle (*caliber* refers to the size of the barrel's interior). Both were muzzled-loaded. Confederate troops used rifles imported from England—mostly the .557 caliber British Enfield. A good marksman could hit a target half a mile away with these guns, and even an average soldier could hit a target at 250 yards.[5]

Although the Springfield and British Enfield were the main rifles used in the Civil War, many other guns appeared late in the war. Christopher Spencer of Connecticut invented a rifle in 1860 that could be loaded with seven cartridges in a hollow in its stock and had a lever that opened the breech. It was one of the first repeaters. It was called the Spencer rifle. Although its existence was well known, it saw limited use in the war.

Another repeater was the .44 caliber Henry rifle, which carried fifteen cartridges. Both the Henry and the Spencer were carbines, which means they had shorter barrels than the standard rifle. The carbine was lighter due to its shorter barrel; as such, it was preferred by the cavalry because full-length rifles were difficult to manage for horse-mounted soldiers. The downside of the shorter barrel was that carbines were usually less accurate and less powerful than larger rifles. Nevertheless, many preferred them because of their light weight and maneuverability. They were particularly good in highly wooded areas where range was not critical.

One of the most interesting rifles of this era was the Sharps, invented by Christian Sharps in 1848. But again it saw only limited use in the war, mostly by marksman, or snipers. One of the main problems was that it was quite expensive—three times the cost of the Springfield. The Sharps got a lot of attention after the 1990 movie *Quigley Down Under*, in which Matt Quigley (the character played by Tom Selleck) had a Sharps with an ultra-long barrel (four inches longer than the standard thirty-inch barrel) that was extremely accurate at a long distance. In the film, Quigley amazed everyone with his ability to hit targets at tremendous distances.

It should be noted also that most of the standard-length rifles had shorter carbines that used the same ammunition, but they had a shorter range and were not as accurate. Even the very accurate Sharps had a carbine model.

Revolvers were, of course, also used extensively during the war, with the colt .44 and .36 being the most popular. The French LeMat revolver was used by Confederate officers. In addition, bayonets, lances, sabers, and swords of various types were used, but in reality they inflicted few casualties.

The other weapon that was used extensively was the cannon. Both smoothbore and rifled cannons saw action. And again, it was the rifled cannon that was most accurate. Projectiles for the cannon were designated by their weight, ranging from twelve pounds up to ninety pounds and more. As in the case of the rifle, the cannon could be breech-loaded or muzzle-loaded. The three main types of artillery were cannons, which fired shells that followed a relatively flat trajectory; mortars, which fired shells into a high, arching path; and howitzers, which were in between. One of the most popular cannons was the smoothbore called the "Napoleon." It was used both by the Union and the Confederates. It was relatively light and portable, and it had a range of about seventeen hundred yards. It was frequently used to shoot canisters and grapeshot, both of which were deadly to an approaching army. Canisters held about eighty-five iron balls, and they disintegrated soon after being fired, spraying the balls over the landscape, causing large numbers of casualties. They were, in effect, like a giant sawed-off shotgun.

THE WAR

Neither side expected the war to last more than six months, and no one expected the ferocity and high numbers of casualties that it caused. By the end of the war over seven hundred thousand soldiers had died, along with a large number of civilians.

The American Civil War began shortly after the election of Abraham Lincoln as president in 1860. Lincoln opposed slavery, and several Southern states feared that he might try to outlaw it. As a result, in 1861 several slave states seceded from the union and formed their own government: the Confederate States of America. Washington did not recognize the secession, and Lincoln was determined to keep the states together.[6]

Fighting began when the Confederacy began shelling Fort Sumter, a garrison in the middle of the harbor of Charleston, South Carolina. After extensive damage, the garrison finally surrendered on April 13, 1861. Lincoln was outraged; he immediately called for volunteers for a Union Army. Soon four more states joined the Confederacy. Within a short time both sides had armies of around one hundred thousand men, but most of them were untrained.

Over the next four years, 237 battles were fought along with many minor skirmishes. And the war became increasingly ferocious as the dead piled up. One of the main reasons for the huge number of casualties was the tactics used by the generals on both sides. Napoleonic tactics had been drilled into the officers at West Point, and they were still used extensively for the majority of the war. As we saw earlier, in the Napoleonic Wars, soldiers arrayed in battle lines marched toward one another with their muskets. They usually fired when there were about a hundred yards apart; at this distance their bullets were barely able to reach the enemy, let alone hit anyone. The order was usually, "Ready. . . . Fire!" rather than "Ready. . . . Aim. . . . Fire!" Their goal was to create a hailstorm of bullets on the enemy, hoping that some of the bullets would inflict damage. Once their single-shot, slow-loading muskets were empty, the soldiers would run at one another with bayonets in hand-to-hand combat. The problem with this was that since Napoleon's time, rifles and cannons had become much more deadly and had a much greater range. Soldiers could now pick out targets and hit them at two hundred yards. So frontal assaults of the type described above became almost suicidal. Nevertheless, a soldier would be quickly branded a coward if he tried to take cover instead of advancing toward the enemy's guns. As a result, soldiers were "mowed down" as they moved toward the enemy. Not until later in the war did generals finally abandon this tactic.

Early in the war President Lincoln ordered a naval blockade of the South in an attempt to stop all trade, particularly the import of weapons from Britain and France. And it was, indeed, quite effective, mainly because the Union had a much better navy than the South had. One of the major things it did was stop the export of cotton, the South's major export, which accounted for a large fraction of its economy. Early in the war the Confederates, led by General Robert E. Lee, won several critical battles. But by the summer of 1862 the Union had destroyed many of the Confederate armies in the West, and it had seriously degraded the Confederate naval forces on the Mississippi River. Then came the Battle of Gettysburg in 1863. Lee was pushing his army north in hopes of a decisive victory that would demoralize the Union. He had approximately seventy-two thousand men; opposing him was a Union force of ninety-four thousand men under the command of General George Meade.

The fighting started on July 1 on the outskirts of the small town of Gettysburg. The Union army was assaulted by Lee's army, sending them retreating to Cemetery Ridge south of the town. On the second day the two armies were again ready for battle, with the Union army laid out in a defensive formation. Late in the afternoon Lee's army charged at several points along the long defensive line and nearly succeeded in breaking through. Lee was now confident; his army had inflicted considerable damage on the Union line, and he was sure victory was at hand. On July 3, he began with an artillery barrage involving 135 cannons, many firing twelve-pound shells, and others firing deadly canisters that could spray troops with lead balls. In addition, he had twelve-pound shells packed with gunpowder that would explode over enemy troops, spraying them with fragments of iron from the shell. Lee's plan was to disable as many of the Union guns as possible, along with soldiers. He knew he would have to devastate and demoralize the Union forces before he could attack.[7]

Soon hundreds of rounds were being fired from the 135 Confederate cannons, and within a short time an equal number of Union cannons joined in. The roar was deafening, but even worse was the smoke from the two rows of cannons. The entire area was soon engulfed in a gray, acrid smoke that stung the soldiers' eyes. The bombardment continued for hours, and to Lee's disappointment, the Union counterbombardment didn't stop or even falter. Nevertheless, by midafternoon Lee decided to go ahead with his attack. Brigadier General Pickett was in charge of the attack; 12,500 Confederate troops lined up for what became known as Pickett's charge. The Confederate soldiers were armed with Enfields, and the Union soldiers carried Springfields. Some of the troops even had the incredibly accurate Sharps. The rifles carried by both armies used Minié balls that had ranges of at least a quarter of a mile. As the Confederates moved forward across the open field, the Union soldiers were waiting for them behind a low stone wall. They let the Confederates move closer and closer. Finally the Union troops let loose, and at the same time Union cannons began firing canisters and exploding shells above them. Some of these shells took out two men at a time, and the Springfields were almost as deadly.

Strangely, the Confederates didn't take cover, but kept coming. The slaughter continued as they approached, and by the time they reach the Union lines, almost half of them—close to six thousand—had already been killed. A brief hand-to-hand battle occurred, but it didn't last long. Within twenty minutes the Confederates retreated, leaving the battleground strewn with bodies.

That night it poured rain, and the battlefield became a field of mud. No one was in the mood for further battle. Seeing the damage to his army, Lee finally decided to retreat. In all, each side suffered incredible casualties—the greatest of the Civil War: twenty-three thousand on the Union side, and about the same number on the Confederate side.

The war continued for another two years after the battle at Gettysburg, but the momentum was now on the side of the Union. In 1864 President Lincoln appointed General Ulysses Grant as chief of all Union forces, and the "limited" war to restore the Union became "total war" to destroy the South and slavery, and to restore the Union. The South fought on valiantly, but it suffered defeat after defeat and was finally brought to bay in April 1865.

THE ROLE OF THE TELEGRAPH

One of the more important devices that came directly from discoveries in physics and was used extensively in the Civil War was the telegraph. It was a communication system that transmitted electrical signals over wires from location to location. These signals could then be translated into a message. President Lincoln used the device extensively throughout the war to keep in touch with his generals and other officers. He was well aware of the value of the telegraph and was the first president to use it to give direct orders on how a war should be conducted.[8]

The roots of the new technology went back to 1823 when the English inventor William Sturgeon (1783–1850) devised the first electromagnet. Faraday had shown that when a current was passed through a coil of wire (what we now call a solenoid) a magnetic field was produced. Sturgeon began by repeating Faraday's experiments, then he went on to wind wire around an iron bar; in particular, he wound eighteen turns of bare copper wire around it and found that it produced a relatively strong magnet, so strong, in fact, that it could lift twenty times its weight of iron. And for the current, he used only a simple single-cell battery.

News of the discovery reached Joseph Henry (1797–1878) in United States, and he decided to do the same experiment with insulated wire. Sturgeon had kept his wires well-separated from one another so they would not short-circuit. But Henry's wires were insulated, so that wasn't a problem; he therefore wound them on top of one another, and he could use hundreds of turns. By 1831 he had produced an electromagnet that could lift over a ton of iron. Then he hooked it into a circuit and showed that it could cause a lever to strike a distant bell. To see how he did this, consider a simple circuit (shown below) that includes an electromagnet. Assume there is a key in the circuit that can be pressed to complete the circuit, but when it is released it snaps back as a result of a spring to break the current so that no current flows. When the key is depressed, the electromagnet in it is actuated and can attract the nearby iron bar. Again, if this bar is part of the circuit, when it is attracted it will break the circuit. Furthermore, if a bell is placed nearby, the bar can strike it when the key is depressed.[9]

Now consider the case of an iron bar that is not part of the circuit. In this case, the electromagnet attracts the bar to it and keeps it there. It releases only when the key is released. So if you press and release the key, the bar will snap back and forth according to the pattern you apply to the key. If the electromagnet is at some distance from the key, a "message" of clicks could be sent from the key to the electromagnet. And what was particularly important was that it traveled at the speed that electricity travels in a wire, which is close to the speed of light.

In 1837 the British physicists William Cooke and Charles Wheatstone patented a device based on this idea, and it is generally considered to be the first electrical telegraph. But it had problems. One of the major ones was that the current in a wire decreases as the wire's length increases because of the resistance of the wire. So a message in the above device couldn't be sent very far. Cooke and Wheatstone did, in fact, invent a device that helped, but it was the American Joseph Henry who improved the device and made distant telegraphy feasible.

Henry devised what is now called the relay. He used a wire for the original current that was short enough so that the signal could still be detected, even if it was quite weak. As a result, the electromagnet could still pull a lightweight key toward it. When it did this, however, it was set up so that it closed a gap in a second circuit that was powered by a battery, and this battery produced a large current in the second circuit. It was not a very long circuit and had little resistance, so the current was stronger. And of particular importance, this secondary current created the same "message" that was flowing in the primary circuit, but it was much stronger. This technique, in fact, could be applied to many additional circuits, so by using relays and batteries at appropriate intervals, you could send a message over a considerable distance. In 1831 Henry sent a message over a distance of a mile, and within a short time this was extended to many miles.

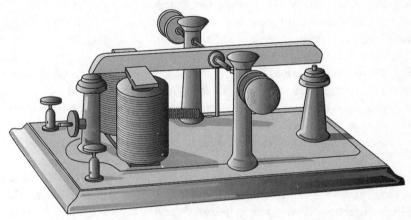

An early telegraph.

But there was still the problem of the message. It was necessary to translate the clicks developed by opening and closing the circuit. Samuel Morse (1791–1872) realized that the clicks could be arranged as a series of dots and dashes that could be sent over the wires, and he set up a code, now referred to as the Morse code. Each letter of the alphabet was coded into a brief series of dots and dashes (A, for example, was a dot and a dash, and B was a dash and three dots). And in 1844 wires were strung between Boston and Washington, DC, and a message was successfully sent between the cities. The message sent was "What hath God wrought," from the Bible's Book of Numbers 23:23.

By the beginning of the Civil War, telegraphy was becoming an important mode of communication across America. The first transcontinental telegraphic system, which ran from California to Washington, was completed in October 1861, about the time that war broke out. And as we saw earlier, Lincoln made extensive use of it. Interestingly, there was not a telegraph in the White House itself, but there was one at the Department of War building, next door, and Lincoln spent many hours in the building. It's estimated that he sent over a thousand messages to his generals and other officers during the war.

THE DYNAMO (GENERATOR)

The Civil War was the first war in which electricity began to play a large role in many different ways. And its impact extended far beyond the telegraph. Initially, most electrical currents were produced by batteries, but batteries are limited in the amount of electrical power they can generate. And for factories and mills, considerable power was needed. The Civil War was, in fact, one of the first truly industrialized wars. Mass-produced weapons, ironclad steamships, large factories producing various goods for the war, railroads, and so on all played important roles. Electricity was central to many of them, and at this point, many of its properties and applications were still not well understood. Furthermore, a cheap source of electricity was still not known.

Nevertheless, the first step had already been taken. Faraday demonstrated electromagnetic induction in 1831 when he showed that a brief current would flow when he moved a magnet in a solenoid. The problem was that the current was too brief, and even if the magnet was moved back and forth, only a fluctuating current could be produced. Faraday decided to look into the possibility of producing a more useful current. To do this he set up a thin copper disk that could be turned on a shaft. The outer rim of this disk would pass between the poles of a strong magnet, and as the disk turned it cut through magnetic lines of force. As a result, a potential difference, or voltage difference, was set up on the disk. The voltage was highest near the rim of the disk, since that was where the disk moved the fastest. Faraday then set up two sliding contacts on the disk, with wires attached to them, one near the edge and one closer to the center. If a galvanometer was placed in the circuit, a current flowed through it, and as long as the disk moved, the current flowed continuously.[10]

But Faraday's disk created only a small voltage difference because it contained only a single current path through the magnet. Soon it was found that much higher voltages could be generated by winding multiple turns of wire into a coil. In 1832 a French instrument maker, Hippolyte Pixii, improved on Faraday's device. He used a permanent magnet that could be rotated by a crank. Placing the magnet so that its poles passed a piece of iron wrapped in insulated wire, Pixii found that the spinning magnet produce a pulse of current in the wire each time a north or south pole passed the coil. But the two poles induced currents in opposite directions. To overcome this, so that both currents were in the same direction, Pixii placed a split metal device, called a commutator, on this shaft (with springs attached to it) that pressed against it.

The result was generally continuous, but not the direct current that we know today. Within a few years, however, a smooth direct current was produced. This was the first dynamo, or simple generator—in short, a device that generated an electrical current as a result of mechanical motion. It meant, however, that you had to have something to push the device around to create circular motion. The steam engine could be used for this, or water in the form of a waterfall, or just flowing water. Electricity, and therefore electrical currents, could therefore be produced if an appropriate outside source of mechanical power was available. The dynamo was the first device that allowed a large amount of electrical power to be generated, such as that needed for a factory.

THE GATLING GUN

Strangely, one of the best "super weapons" of the Civil War saw little action. The Gatling gun was designed by Dr. Richard Gatling in 1861 and patented in November 1862, but the army appeared to have little interest in it at first. Also strange is the fact that Gatling abhorred war and hoped that his weapon would overcome the need for large numbers of soldiers on the battlefield. Even more so, he hoped it would show how gruesome and terrible war could be, which might convince nations to think twice before they went to war.[11]

The Gatling gun used multiple rotating barrels to fire two hundred bullets per minute. The barrels, six in all, were mounted around a central shaft, and the entire assembly could be rotated with a hand crank. Each barrel fired a single shot as it rotated to a certain point. The shells consisted of steel cylinders containing black powder and a percussion cap. Compared to other attempts to increase firepower, the shells used in the Gatling gun were gravity fed into the breach from a hopper on the top of the gun. After each bullet was fired the empty cartridge was ejected and a new round was loaded. One of the major problems in earlier attempts at such a gun was overheating of the barrels. In this case the barrel was allowed to cool as it rotated; in addition, in the earliest models, fibrous matting that had been soaked with water was stuffed between the barrels.

Gatling demonstrated his new weapon to the Union army in December 1862, several months before the Battle of Gettysburg, but the army was slow to accept it, perhaps for the best, since it would soon have become a major killing weapon.

THE WAR AT SEA

While war raged on land, war was also taking place on the high seas and in the bays along the Gulf of Mexico, and even up the larger rivers such as the Mississippi. Soon after the war began, Lincoln ordered a blockade of the seaports in the South, and it was, indeed, a smart move. The South had limited resources and had hoped to get support, or at least supplies, from Europe, and with the blockade on, they didn't get much of either. The blockade was particularly effective because the navy, as limited as it was at the time, remained loyal to the Union. At the time, in fact, it had only wooden ships, which would soon become so vulnerable to gunfire that they were of little use, unless guarded by an "ironclad."

As guns got bigger and bigger, it soon became obvious that wooden ships were sitting ducks. Something would have to be done. Iron or steel plating was placed over the wood at first, but it soon became apparent that it would be better to make the entire hull out of metal. Such ships became known as ironclads.

Early on most ships were propelled by a giant paddlewheel, which in turn was powered by a steam engine. But paddle wheelers were large, cumbersome, inefficient, and vulnerable. One well-directed artillery shell and the ship would be out of commission. Engineers were therefore looking for something better for propulsion, and the obvious thing was some sort of "screw" device. Centuries earlier Archimedes had used a screw-like propeller for lifting water for irrigation, and the Egyptians had used a similar devise for years to irrigate their lands. Furthermore, Leonardo da Vinci had used the same principle in his design for a simple helicopter. It was obvious it could move water and exert a force against water. One of the first to propose that such a device could be used to propel a boat was James Watt, but strangely there's little evidence that he suggested using it with his steam engine.

The first "screw" propellers were, indeed, in the form of a long screw. But in 1835 Francis Smith made an important discovery. He was experimenting with long-screw propellers when a large section of the screw broke off. To his surprise, the remaining piece of the propeller seemed to work even better than the long screw. This was the beginning of the shorter, modern propeller that we know today. The man who perfected the design was the engineer John Ericsson of Sweden. By 1836 he had added larger blades and made it much more efficient. He worked in England for a while but came to the United States a few years before the Civil War, and his talent soon came to the attention of the naval officer Captain Robert Stockton. Stockton was ambitious and was determined to modernize the navy with armored steamships and much larger guns. With the help of Ericsson, he designed and built one of the most formidable warships of the day, naming it the *Princeton*, after his hometown. Its guns were in turrets, the two largest of which had twelve-inch barrels that fired 212-pound cannon-balls. In addition it had twelve forty-two-pound guns, and it was powered by Ericsson's new propeller.[12]

In 1844 it was paraded out before President Tyler and a large audience in Washington, DC. Stockton, eager to show off its guns, ordered a demonstration. As the third shot was fired, the large gun exploded, spraying the attending crowd with fragments of iron. The secretary of state, the secretary of the navy, and several other officials were killed. It was a tremendous blow to both Stockton and Ericsson, who had helped in the design.

When the Civil War broke out, however, Eriksson was employed by the Union to design a new and even better warship. When completed, it was called the USS *Monitor*. It was heavily armored in iron and was 179 feet long and powered with a steam engine that drove a nine-foot propeller. It looked rather strange in that its deck was only eighteen inches above the water, but this made it an extremely poor target for enemy ships.

In the meantime the Confederate navy had also built an ironclad, which they named the CSS *Virginia*; it was the pride of the South. In March 1822, the two ships met at Hampton Roads, Virginia. The *Virginia* attacked the Union blockade squadron at Hampton Roads and destroyed two small frigates. Early in the battle the large frigate *Minnesota* had run aground while attempting to engage the *Virginia*. But it got dark before the *Virginia* could finish off the *Minnesota*, so early the next morning it returned, but in the meantime the Union navy had brought the *Monitor* in, and it was waiting for the *Virginia*. The two ironclads blasted at one another with their guns but did little damage, then the *Virginia* tried to ram the *Monitor*, but this also did little damage. The two ships continued to battle for hours, but in the end it was a draw. However, the *Monitor* had stopped the *Virginia* from destroying the *Minnesota* and several other ships.

Pleased with the performance of the *Monitor*, the Union soon built an entire fleet of ships modeled on it. They also built a fleet of smaller ironclads referred to as the "City" class, which were used in the west—the bays of the Gulf of Mexico and the larger rivers such as the Mississippi.

The Confederate navy also built several smaller ships, but it was soon obvious that they could not keep up with the Union navy, and they could do little on top of the water to stop the blockade.

PHYSICS OF THE PROPELLER

Propellers of the time had two or more blades that were attached to a rotating shaft. As the propeller turned, it transmitted power to the boat by converting rotational motion into forward thrust. Basically, a pressure difference was produced on opposite sides of the blade, with the pressure on the rear surface greater than that on the front, and this differential forced the ship forward. In effect, the blades imparted momentum to the water, which created a force on the ship.

Propellers can turn clockwise or counterclockwise, according to the design of the blades. The force on the blade depends on its area (A), the fluid density (ϱ), the velocity (v), and the angle of the blade to the fluid flow (α).

One way to look at the propeller is to compare it to a screw. You know that to screw it into a wall, you apply torque to the head of the screw. The helical thread of the screw converts this torque into a "pushing" force that drives the screw into the wall.

Basically, a propeller is a machine that moves the ship through the water as it is turned. Machines, as we saw earlier, are devices that multiply or transform forces. So a propeller is a machine that moves the ship forward by pushing water backward, where the force on the backward-moving water is equal to the force on the forward-moving ship, according to Newton's third law. Also, since the force is a result of a change in motion, a propeller gives a ship forward momentum by giving the water backward momentum.

"DAMN THE TORPEDOES"

When the Confederate navy finally realized it was no match for the Union navy it decided to fight the embargo in other ways. And two of the most effective things they used were torpedoes and submarines. Indeed, the Confederates sank twenty-two Union ships and damaged twelve others using torpedoes, while losing only six ships to the Union navy in return. These "torpedoes," however, are not what we usually think of as torpedoes, even though they were referred to by that name at the time. They were what we might call "mines" today.

Two types of torpedoes were used extensively: the spar torpedo, in which an explosive device was mounted at the end of a long spar (up to thirty feet long). It was usually mounted at the bow of an attacking vessel. When driven against an enemy ship it exploded. The only problem is that it frequently did considerable damage to the vessel carrying it. Torpedoes were also towed on long ropes or lines behind a vessel, usually at an angle of about forty-five degrees. Using

THE AMERICAN CIVIL WAR

an appropriate maneuver they could also be projected at an enemy ship. And, of course, many "torpedoes" were merely set in the water. They could be detonated electrically by an operator on the shore or by some type of percussion cap.

One of the most famous battles in which torpedoes played in important role was the Battle of Mobile Bay, Alabama. It occurred in August 1864. On the Confederate side, guarding the bay was Admiral Franklin Buchanan, a veteran of numerous battles at sea. His flagship, the *Tennessee*, was an ironclad modeled after the *Virginia*. Outside the bay was Union admiral David Farragut, who had four ironclads modeled after the *Monitor*, along with several wooden ships. And he faced more than just the *Tennessee* and a few smaller ironclads. Two forts were guarding the entrance to the bay: Fort Morgan, with several large guns, and Fort Gaines. But the biggest fear for Farragut was the "torpedoes" that were floating throughout the bay. The only way through them was a narrow path directly under the guns of Fort Morgan.[13]

Farragut planned to attack using a formation of two columns of ships. One, which would pass close to Fort Morgan, consisted of four well-protected iron-clads similar to the *Monitor*. The second column consisted of four wooden war-ships, lashed together for safety. If one was hit, it was less likely to sink. Farragut was in the second of these, in the *Hartford*; ahead of him was the *Brooklyn*. On August 5, Farragut's armada approached the bay, and as they closed in, the guns at Fort Morgan began firing at them. The Union ships began firing back, but Farragut was not looking for an extended battle with them; he planned on rushing by them into the bay.

As the two columns approached the bay entrance, the captain of the lead ironclad, the *Tecumseh*, spotted Buchanan's *Tennessee*. It was a major danger to the column of wooden warships. But as he moved to intercept it, he forced the wooden ships into the minefield. When the captain of the *Brooklyn* saw the mines ahead he ordered his ship to stop. But directly behind him was Farragut in the *Hartford*. Annoyed, Farragut sent a message (flag message) to the captain of the *Brooklyn* to continue on. Confusion reigned, as both ships were under heavy fire. All at once an explosion rocked both ships. The lead ironclad, the *Tecumseh*, had hit a mine, and within seconds it was at the bottom of the bay.

Confusion continued, as the *Brooklyn* stopped again. It appeared as if the entire column of ships was going to collide into one another. Farragut issued an order that the line to his ship be cut, and he pulled out and began steaming past the *Brooklyn*. The captain of the *Brooklyn* yelled at him, "There are torpedoes directly ahead," to which Farragut replied, "Damn the torpedoes . . . full speed ahead!" (a phrase that has become legendary). As the *Hartford* steamed ahead it struck several mines, but luck was with them. None of the mines exploded.

Meanwhile, Buchanan, who was aboard the *Tennessee*, watched in astonishment as all the Union vessels passed safely into the bay. He ordered the *Tennessee* to steam directly at the *Hartford*, which was now leading the Union ships. He planned to ram it, but the *Tennessee* was large and slow, and the *Hartford* easily eluded it as gunners from the two ships fired at one another. The *Tennessee* then made runs at several other ships, hoping to ram them, but it did little damage, so Buchanan broke off and returned to Fort Morgan.

But the fight was far from over. After inspecting his ship for damage, Buchanan ordered it out to sea again. And again the two ships—the *Hartford* (at ten knots) and the *Tennessee* (at four knots)—steamed directly toward one another. It looked like they were going to collide when, at the last moment, the *Tennessee* veered slightly. As they passed one another at point blank range, sailors on both ships fired with muskets and pistols.

Once the *Tennessee* was past the *Hartford*, however, it was surrounded by Union warships, all firing at it at the same time. And with the range being exceedingly close, they did a tremendous amount of damage. Furthermore, the gun port for one of *Tennessee*'s guns jammed, and some of its other guns misfired. Then one of the incoming shells took out the *Tennessee*'s steering, and it could no longer maneuver. Finally, Buchanan himself was struck by flying debris. There was nothing left for him to do but surrender, and he did.

SUBMARINES

The first submarines also saw action during the Civil War. Actually, the first submarine had been built many years before the war, in 1776 in England. It was a one-man, hand-cranked machine. And the American inventor Robert Fulton had constructed a submarine for the French navy.

As we saw earlier, the Confederates soon realized that they were badly outgunned on the surface, so much of their effort went into the region under the surface—in particular, they used submarines. In 1862 they built the first of several submarines, all of which were called "David." (The name no doubt came from the biblical story of David challenging the giant Goliath.) It was steam driven, and, as such, it needed a smoke stack, and since both the smokestack and the breathing tube had to penetrate above the surface, it was quite limited. Its major weapon was a spar torpedo, which was mounted on the bow.[14]

Within a short time Horace Hurley and two partners launched the *Pioneer*, and in 1862 they launched *Pioneer II*. By now they were experimenting with electric engines, but the only electrical engines that were being manufactured were in the North. They attempted to smuggle some in but failed. The following year, the much larger *Hurley* was built (named for the maker). It was forty feet long and was about four feet in diameter, and it had an eight-man crew that turned a hand-cranked propeller. The hand crank was no doubt used in an effort to keep the machine as silent and undetectable as possible. The *Hurley* had a spar torpedo, and there is evidence that it was used several times in battles, and no doubt several crews died in their attempts to do so. Nevertheless, it is the only submarine that managed to sink an ironclad during the war. In 1864 it sank the Union sloop *Housatonic*. Unfortunately, it did not survive the attack and was never seen again. In 1995, however, its remnants were found off the coast of South Carolina, and it was raised in 2000. There is some evidence that it was only twenty feet away from the Housatonic when it exploded, and the concussion likely disabled it.

The Union was not very active in the production of submarines, but they made one that they called the *Intelligent Whale*, but it never saw action. Interestingly, several private ventures in both the North and the South also attempted to build submarines, but little is known about most of them. In all there may have been twenty submarines built during the Civil War, with most of them seeing no action. But the experimentation and innovation that went into them soon led to much better submarines. In particular, airlocks, compression air ballast tanks, electric motors, periscopes, and air purification systems were developed.

BALLOONS

Hot-air and hydrogen-filled balloons were used for surveillance by both the Union and Confederate armies during the war. But they were used more extensively and effectively by the Union army. In 1861 Lincoln gave orders to form a balloon corps with Thaddeus Love in command. And indeed, in several battles the information obtained using them was of considerable value. During the Seven Days Campaign of 1862, for example, Union balloons stationed seven miles from Richmond could easily observe troop movements within the city. The largest balloons (called the *Integral* and the *Union*) could carry five people and had a capacity of thirty-two thousand cubic feet. Hydrogen gas was used in most of the early balloons; it was generated from water using portable generators.[15]

Almost all balloons were tethered to the ground by a long line, but they were able to climb to almost five thousand feet in the air. And although Union balloons were shot at extensively by Confederate cannons, they were generally too high to be in range, and none were ever knocked down. Most of the larger balloons also had telegraphic equipment to transmit information to the appropriate people below.

How did these balloons work? For a balloon to rise, there has to be a force on it, and the only force readily available is a buoyant force. The Greek mathematician Archimedes was the first to understand this force, and as we saw earlier, it is now known as Archimedes principle. It states that *any body completely or partially submerged in a fluid (in this case the fluid is air) is buoyed up by a force equal to the weight of the fluid displaced by the body.*

In the case of a balloon, a buoyant force (B), which is equal to weight of the air displaced, acts upward, and a gravitational force (W) acts downward. To see how this causes a balloon to rise, let's begin with the density of the gas (ρ) and the volume (V) of the balloon. The mass of the displaced air is going to be ρV, and for its weight we need to multiply by g (the force of gravity). So the buoyant force B is ρVg. Now we need W, the weight of the balloon. For it we need the mass inside it, and this is the density of the gas inside it (D) times its volume (V), or DV, so $W = DVg$. The total upward force is therefore $B - W = \rho Vg - DVg$. If the upward force is positive, the balloon rises. Since hydrogen is less dense than air, the balloon will rise if filled with hydrogen (or, for that matter, any gas that is less dense than air). Also, if we heat the air inside the balloon, the molecules move farther apart and its density decreases. And again the net force is upward. This is the principle of the hot-air balloon.

WHERE DOES THE BULLET GO?

Ballistics of Rifle Bullets and Cannon Shells

In an earlier chapter we were introduced to the problems related to the accuracy of rifles and cannons. For many years, in fact, gunners had no idea what the trajectory of a bullet looked like. Tartaglia made some important advances, and Galileo clarified many of the problems related to gravity, but it was Newton who finally showed why and how gravity was involved. In this chapter we will look much more closely at the problem, and although we have discussed only muskets, rifles, and cannons up to the time of the Civil War, we will also deal with more modern rifles and cannons in this chapter.

The study of ballistics is concerned with the motion of projectiles, but it is also concerned with what happens to the projectile, or bullet, inside the gun, and also what happens when it hits a target. In all, there are four basic areas of study within this topic:

- Interior ballistics
- Transitional ballistics
- Exterior ballistics
- Terminal ballistics

Interior ballistics is concerned with what happens between the firing of the cartridge and the exit of the projectile from the muzzle. Transitional ballistics, which is also known as intermediate ballistics, is a study of the projectile from the time it leaves the muzzle until the pressure behind the projectile is equalized (in other words, when the air pressure behind the projectile is equal to the surrounding air pressure). Exterior ballistics deals with the projectile while it is in flight under the influence of gravity. Terminal ballistics is the study of what happens to the projectile after it hits the target. I will talk mostly about rifles, but much of what I say also applies to cannons.

INTERNAL BALLISTICS

Internal ballistics depends on what happens inside the breach and barrel of the gun, so it is the best place to begin our discussion. Of particular importance, the trajectory depends, to a large degree, on what is referred to as the muzzle velocity, which is the speed of the bullet as it leaves the end of the barrel. So let's look at how muzzle velocity develops. Two separate events are critical here: ignition of the gunpowder, and expansion of the gases that develop as a result of the initial explosion. Ignition occurs when the firing pin hits the percussion cap (or primer) and causes it to explode. This, in turn, ignites the gunpowder in the cartridge. As the gunpowder explodes, the gases produced by the explosion are trapped behind the projectile. They are at high temperature and therefore create pressure that accelerates the projectile down the barrel. It is particularly important that the burn time is less than the time it takes the projectile to reach the end of the barrel. If not, powder will come out of the barrel still burning, and this would create a dangerous situation.[1]

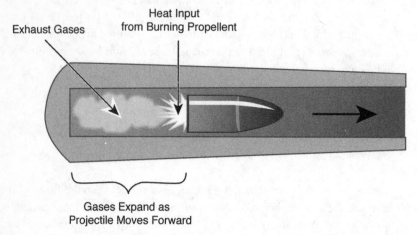

Expanding gases behind a bullet in a barrel.

As the gas expands, it cools according to the basic gas law discovered by Jacque Charles in the late 1800s, which is now referred to as Charles's law (it is sometimes referred to as Gay-Lussac's law, since it was discovered by Joseph Louis Gay-Lussac about the same time).[2] This law tells us that pressure times volume is proportional to temperature, and therefore a sudden increase in temperature will create an increase in volume that will impart a higher pressure on the projectile. Inside the barrel the pressure does, indeed, increase initially, but as the projectile moves forward, the pressure decreases. A plot of pressure versus distance along the barrel is given below. We see that the pressure reaches a peak rather quickly before dropping rapidly. The maximum pressure is important because it occurs mainly in the breach area of the gun, and this area has to be able to withstand the pressure. This is why the steel of guns and cannons is thickest in this area. The rearward push on the gun's bolt, or breach, which results from the explosion, is referred to as the bolt thrust. It depends on both the chamber pressure and the diameter of the cartridge case. And it's important that the design of the gun is sufficient to withstand the bolt thrust. The pressure in this area (the chamber) is usually measured in pounds per square inch (psi), or in the metric system it is measured in kilograms per square centimeter (kg/cm^2). Typical pressures in this area depend on the type of gun, but for rifles it is approximately fifty thousand pounds per square inch.

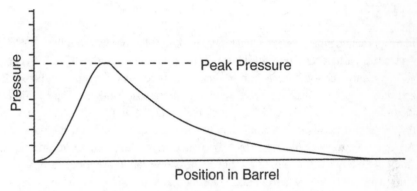

A plot of pressure versus distance along the barrel.

It's fairly obvious that the longer the projectile is accelerated by the explosive force, the faster it goes. The law controlling this is Newton's second law, which states that force is equal to mass times acceleration (F = ma). And this acceleration continues as long as the projectile is in the barrel (and slightly beyond, as we will see later). So the longer the barrel, the greater the velocity of the projectile when it exits; this is called the muzzle velocity. And although longer barrels provide greater velocity, there is a limit. Very long barrels are hard to handle, and the weight of the gun also increases with barrel length.

The length of the barrel is important in another way. As the hot gas explodes down the barrel, it decreases in pressure, and ideally it should not be much higher than atmospheric pressure when it reaches the end of the gun. In practice, however, it is usually much higher. As a result, it creates a shockwave as it hits the air. The problems related to this will be discussed in the section on transitional ballistics.

In addition to barrel length, the muzzle velocity also depends on the mass, or weight, of the projectile for a given amount of explosive; a lighter projectile will emerge with a greater muzzle velocity. The type of powder in the barrel is also an important factor; different types of powder have different propellant energies. How much powder should be used is also a factor, but for a given caliber, there is a limit on the amount of propellant.

Another important question is: What is the maximum velocity that a projectile can be pushed to without extreme danger to the gunner? In rifles this is about four thousand feet per second. Large-caliber guns and cannons can safely be pushed to about six thousand feet per second.

RECOIL

It's important to note, however, that higher muzzle velocities have another problem. As we saw earlier, Newton's third law tells us that for every action there is an equal and opposite reaction. The action in this case is the force pushing the projectile and the hot gases out of the barrel, creating the muzzle velocity. The reaction is therefore opposite to this, and the gunner feels it as the gun's recoil. The direction of the recoil force is directly opposite to the force that pushes on the projectile. Anyone who has ever shot a gun has experienced recoil, and at times it can be quite strong. Directly related to the third law is the conservation of momentum. We can write it as $mv = MV$, where m and v are the mass and velocity of the projectile, and M is the mass of the gun (or the gun and gunner) and V is the recoil velocity. The mass of the gun is much greater than the mass of the projectile, but the muzzle velocity of the gun (v) is extremely large, so without constraint, the gun would achieve a relatively high velocity. But it is constrained by your shoulder, and this creates a relatively large force on it, which slows its velocity very rapidly. Gunners are, in fact, told to hold the gun snugly against their shoulders so that M includes not only the mass of the gun, but also the mass of the gunner.[3]

Also, in movies and TV, you've likely noticed that the actors always hold their revolvers with two hands in an effort to steady them. One of the reasons for this is that recoil tends to cause the revolver to turn upward. This is caused by an upward torque that occurs because the recoil force is pointed along the barrel of the revolver, but the revolver is held by a lever, namely your arm and shoulder. Since this force is at an angle to the level arm, it creates it torque, causing the barrel to move back and turn up at the same time.

One of the major ways to reduce the effects of recoil is to have a recoil pad at the end of the gunstock.

TRANSITIONAL BALLISTICS AND THE SONIC BOOM

Transitional ballistics is sometimes referred to as intermediate ballistics because it is concerned with the projectile behavior in the time between internal and external ballistics; in other words, the brief time from when the projectile leaves the barrel until the pressure behind it reaches the pressure of the surrounding air. When the bullet reaches the end of the barrel, the gases behind it are frequently at a pressure several hundred times that of atmospheric pressure. When the bullet flies free of the barrel, however, the gases behind it are free to expand and move outward in all directions. This sudden expansion causes a loud explosive sound, namely the boom that you hear when a gun is fired. It is also sometimes accompanied by a flash as the gases combine with the oxygen of air.[4]

This is the first boom you hear when a gun is fired, but in many cases there is also a second boom, referred to as a sonic boom (it will be discussed a little later in the section). Earlier I mentioned that the bullet accelerates as it moves up the barrel because it is being pushed by the expanding gases, but once it leaves the barrel it has a uniform horizontal velocity. This is not exactly true. There is still a force on the bullet for a short time after it leaves the barrel due to the expanding gases behind it. This is one reason why the "muzzle velocity" of a rifle is not measured at the end of the barrel, but several feet in front of it.

In the design of a gun it is important to make sure that the gases that expand behind the bullet as it leaves the gun do not disturb the bullet from its path. If it is somehow pushed off to one side, the accuracy of the rifle would be compromised. So the gun has to be designed so that this doesn't occur; in short, the designers have to make sure the gases expand symmetrically around the base of the bullet.

In the case of military weapons—particularly sniper rifles—it is important to decrease the sound and flash from the rifle as much as possible so that the position of the sniper is not given away. This is done by flash and sound suppressors; in both cases devices are used to change the flow of the escaping gases. In the case of flash suppressors, turbulence in the escaping gas is introduced in an attempt to reduce the combustion efficiency of the flash. In the case of sound suppressors the gas is allowed to cool so the velocity at which it exits the barrel is reduced; this prevents a shockwave from forming. Unfortunately, suppressors are bulky and heavy, and they are not used extensively.

But even if they reduce the shockwave from the exiting gases, there is still a sonic shockwave that can easily be heard. Let's look at this wave. It's well known that any object traveling through air at a velocity greater than that of sound will cause a sonic boom. This applies to most bullets, and, as it turns out, no type of silencer can dampen a sonic boom, since the shockwave that causes the boom travels along with the bullet.

For a supersonic shockwave, the bullet has to exceed the speed of sound, which is about 1,100 feet per second (or approximately 750 miles per hour) depending on the air pressure and a number of other factors. About one-half of all pistol ammunition is supersonic, as is virtually all rifle ammunition, and most shells shot from cannons are also supersonic. Tank guns, as I mentioned earlier, have velocities up to six thousand feet per second, which is many times the velocity of sound.

To understand the sonic boom better, let's look at how it is created. It's well known that when an object creates a sound, a wave travels outward from the object at the speed of sound. If you look closely at this wave you see that it consists of a series of compressions and rarefactions. The compressions occur because the molecules of air are pushed together in certain regions, and the rarefactions are caused because the waves spread out in other regions. This means that a wave that is uniform in all directions passes outward from the source. But when you move the object that is creating the wave, the wave pattern around it changes. The compressions get closer together in the direction that the object is traveling and farther apart in the opposite direction. Furthermore, as the object moves faster, the waves in the forward direction start merging into one another, and at the speed of sound they merge completely together.

At this point the pressure on the nose of the bullet is much greater than it is on the rear of the bullet. But sound in air can only move at approximately 1,100 feet per second, and a bullet can move at any speed; in particular, it can move at speeds greater than the speed of sound. Because of this, when the bullet breaks, or passes through the sound barrier, it creates compressions faster than the compressions themselves can move away, so they just pile up on one another. As these compressions are brought together, they do not form a smooth progression from compression to rarefaction, as they do in ordinary sound waves. Instead, there is a sharp dividing line between a volume of strong compressions and the normal atmosphere around the wave. As a result, the strong compressions stream backward in a cone-shaped band. When this cone passes an observer on the ground, he or she experiences a sudden difference in pressure as it moves by, which he or she interprets as a sonic boom. In many ways it is quite similar to the crack a bullwhip makes.

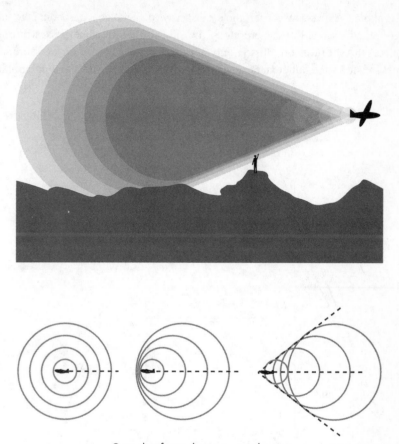

Cone that forms during a sonic boom.

EXTERNAL BALLISTICS

External ballistics deals with the behavior of the projectile in flight from the time it is just beyond the end of the barrel until it hits the target. Galileo realized that two separate motions were involved: a horizontal motion parallel to the ground, and a vertical motion. And although there were two motions going on at the same time, they could be dealt with separately. The horizontal motion was the horizontal component of the muzzle velocity, and it had a constant velocity. The vertical motion was free fall due to gravity, and it was therefore a constant acceleration of 32 ft/sec^2, which is the acceleration of gravity. Galileo also showed that the overall trajectory when you combined the two motions was a parabola. (As we saw earlier, the easiest way to visualize a parabola is to take a cone and slice it somewhere along the side so that the slice passes through the base.) As it turns out, this is only approximate because of air pressure. Air pressure slows the bullet and causes the trajectory to deviate from a parabola.[5]

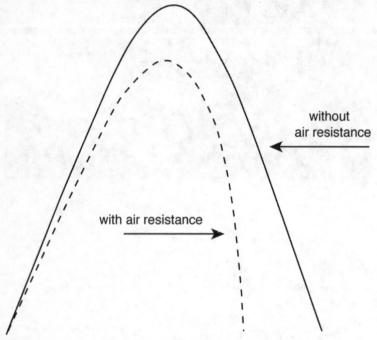

Path of a bullet with and without air resistance.

One of the things that's easy to show is that a bullet drops in the same way that something drops if you hold it above the ground and let it go. A demonstration of this is frequently used in physics classes; it has a simple projector (a gun) that throws an object straight out from it, parallel to the ground, and releases another object at the same time that falls directly downward. The first object traces out a much longer path, but the two objects hit the ground at the same time.

Let's look at the air pressure around the bullet in more detail. It creates a force called drag, which acts in a direction directly opposite to the direction the bullet is traveling. And interestingly, it is a much greater force than gravity (fifty to one hundred times greater), but gravity is still the main force that determines the bullet's trajectory. In practice, the shape of the bullet has some effect on its path, but in a first approximation we can assume gravity is acting on the bullet at its center of gravity. Basically this is just the "balance point" of the bullet.

The drag caused by air resistance actually depends on several things, such as the bullet's speed, its shape, the density of the air that it is passing through, and the air temperature. In practice, calculating drag is usually a difficult problem. Furthermore, there's a serious problem at the speed of sound, or, more specifically, when the bullet passes through the speed of sound. For this reason it is best to deal with four separate regions:

- Below the speed of sound, up to about 1,000 ft/sec.
- Just below the speed of sound, from 1,000 ft/sec to 1,200 ft/sec.
- The peak drag region, which is 1,200 ft/sec to 1,400 ft/sec.
- The supersonic region, which is above 1,400 ft/sec.

If we refer to drag as D we can see how it varies in the three regions by plotting it against the velocity of the bullet.

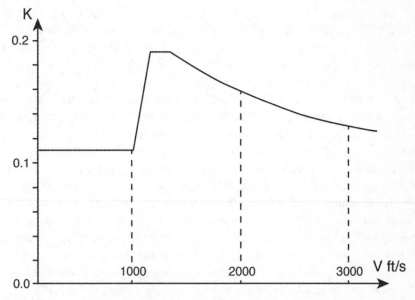

A plot of k (drag force/velocity^2) versus velocity.

The ballistic coefficient (BC) is a term that denotes the rate at which the bullet slows down. In conjunction with the muzzle velocity of the bullet, it gives us a good approximation of the bullet's trajectory. The ballistic coefficient (BC) is defined in terms of what is called the sectional density (SD) and a form factor (FF). The sectional density is the mass of the bullet divided by its caliber squared. The form factor is a measure of the aerodynamic efficiency of the bullet, which depends on it shape, so it is more difficult to determine. In terms of SD and FF, the ballistic coefficient is BC = SD/FF. So if we know the ballistic coefficient, the muzzle velocity, and the angle at which the gun was aimed, we can plot the bullet's trajectory. In practice, however, you need tables giving information about the bullet, so I won't get into it in detail. But we can state the following:

- Bullets with high BCs are the most aerodynamic and bullets with low BCs are the least aerodynamic.
- High BCs are desirable because they give a flatter trajectory for a given distance.
- Bullets with high BCs get to the target faster, so their trajectories are less likely to be altered by wind or other variables.

There are other things that also affect the flight of the bullet. Wind velocity can have a serious effect, particularly if it is at right angles to the direction of flight. In addition, the wind velocity frequently varies over the distance of the flight. Yaw, which is a consequence of the spin of the bullet, can also be a problem; it is a rotation of the nose of the bullet away from the line of flight. A similar effect, called precession, also occurs in the case of a spinning object such as a bullet. It is a rotation around the center of gravity of the bullet. It is easily seen in a gyroscope. Finally, there is something that is only important in very long-range shells. It is referred to as the Coriolis force, and it is created by the rotation of the earth. In effect, the earth rotates under the shell as it moves in flight, but from the perspective of an observer on the ground it appears as if the shell is moving away from its intended trajectory.

Another thing that is particularly important in the case of guns is their maximum range. In other words, at what angle do you aim a gun for maximum range? Galileo showed that in an ideal case, where there is no air resistance, maximum range is achieved when the gun is aimed at an angle of 45 degrees to the ground. But, of course, air pressure changes things quite dramatically. We now know that rifle bullets achieve the greatest range for an angle of 30 to 35 degrees. High-velocity, large-caliber artillery, on the other hand, achieve the greatest range at an angle of 55 degrees. The maximum range of a bullet, however, is not equal to its effective range. The effective range is the range that produces reasonable damage. In general, the greater its mass, the closer the effective range is to the bullet's maximum range. Light bullets, like .22 caliber bullets, for example, have a maximum range of nearly a mile but an effective range of only about a hundred yards.

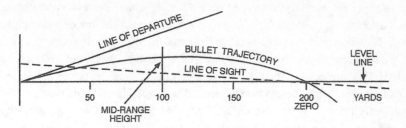

Trajectory of a bullet.

STABILITY OF THE BULLET

As we saw earlier, the main thing that stabilizes a bullet is spin, and this spin is created by the "rifled" interior of the gun's barrel. A rifled barrel has spiraled or helical grooves down its length. The bullet is forced into these grooves, which creates spin along its long axis. When the bullet emerges from the barrel, it behaves like a gyroscope. In particular, it has the stability of a gyroscope. If you have ever played with a gyroscope, you know that it takes considerable force to move it out of the direction it is spinning. This is what gives the bullet its stability and its increased range; without stability the bullet would tumble while in flight, and air pressure would act on it much more strongly.

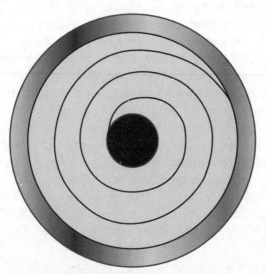

Spiraling grooves in the barrel of a rifle.

Rifling is generally quantified by twist rate, which expresses the distance the bullet travels down the barrel while it makes one revolution, or one complete turn. The shorter the twist distance, the greater the spin rate. If you look closely at the spiraling inside the barrel, you will see that it is a series of grooves with relatively sharp edges. The spaces that are cut out along the barrel are, indeed, called grooves, and the regions that are left are called lands. This type of rifling is usually referred to as conventional riffling, but there is also another type. In this type the entire barrel is cut in the shape of a polygon (e.g., a hexagon), with the polygon shape given a twist as it goes down the barrel. It is referred to as polygonal rifling. In the case of larger shells, such as those shot from ship guns and tanks, the shell is equipped with fins that ride in grooves as they pass through the barrel.

In most cases the spiral rifling is encountered by the shell almost immediately after it leaves the chamber; in some cases, however, the spin is increased gradually. This is referred to as gain-twist. In this case there is no rifling from the chamber to the "throat" of the gun. When the bullet first leaves its cartridge, therefore, it is not spinning. However, it encounters rifling after it has traveled a short distance, and in most cases the rifling increases its twist rate gradually. This allows the projectile to spread out its increase in torque over a larger distance.

The number of grooves in the barrel can vary, as can their shape and depth. Furthermore, the twist direction can be either clockwise or counterclockwise, and the twist rate can vary depending on the bullet shape, weight, and length. In breach-loaded guns the projectile is placed in the chamber. When it is fired, "seating" occurs in the throat. The throat is usually slightly larger than the bullet, so when the bullet is fired it expands under the pressure of the gas behind it until its diameter matches that of the interior of the barrel. The enlarged bullet then travels down the throat to the rifling, where it is "ingrained"; in other words, grooves are cut into it. As a result of these grooves, the bullet begins to spin.

The twist rate given to a particular bullet is critical. First of all, it has to be sufficient to stabilize the bullet, but it should be no greater than this, and the ideal twist rate depends on the bullet's weight, length, and overall shape. Twist rates can vary considerably; older guns, for example, frequently had twist rates as low as one in seventy inches. More modern guns have much higher twist rates, such as one in twelve inches, or even one in ten inches. In general, rifles have much greater twist rates than pistols. The twist rate (T) is frequently expressed as $T = L/D$, where L is the length for one revolution, and D is the barrel diameter.

If the twist rate is too small, the bullet will yaw (move back and forth along the direction of flight), and if this happens, it will eventually begin to tumble and lose its accuracy. A twist rate that is too low can also cause a bullet to precess around its center of gravity. As we saw earlier, this is a motion you can easily see in a gyroscope.

On the other hand, the twist rate can also be too high. In many spinning objects there is an outside force usually called the centrifugal force (this is actually an erroneous term), and the faster the spin, the greater this force. Countering it is a cohesive force that is holding the bullet together. When the centrifugal force becomes greater than the cohesive force, the shell disintegrates and flies apart. In theory, a bullet can have a spin rate up to about three hundred thousand revolutions per minute. Most bullets, however, have spin rates much smaller than this, typically in the range twenty to thirty thousand revolutions per minute.

TERMINAL BALLISTICS

Terminal ballistics is a study of what happens to the bullet or projectile after it hits the target. It's obvious that its velocity will change rapidly. It may be stopped by the target, or it may pass through it. There are two ways in physics to deal with this case; they are referred to as the force or momentum picture, and the energy picture. In the force picture we deal with forces (or momenta), so we use Newton's third law, which says that for every action force there is an equal and opposite reaction force. In this case we are concerned with the force applied to the target, or the momentum delivered to it. In the other view, the energy picture, we are concerned with the kinetic, potential, and any other types of energy that might be involved. In most cases the energy picture is the easier of the two to use, and the major reason for this is that conservation of energy states that energy cannot be created nor destroyed; it can only be changed from one type into another. So for a given problem all you have to do is look at each type of energy involved to make sure that everything adds up. In the case of a bullet fired from a gun the chemical energy of the bullet is transformed immediately to gas pressure and heat energy in the barrel. This energy is then transferred to the kinetic energy of the bullet's motion plus sound energy. And some energy is lost to air resistance. It's the kinetic energy that the bullet finally has just before it hits a target that is important.[6]

Several things can happen when a bullet hits a target. If the bullet stops within the target, it transfers all its kinetic energy to the target, and at the same time it transfers its momentum to the target. It may also pass through the target and emerge at the other side. In this case the bullet transfers some of its kinetic energy and some of its momentum to the target. Finally, it is possible that in the case of a well-armored target, the bullet could bounce back. In this case it delivers all its kinetic energy to the target, but the target actually receives more momentum than the bullet initially had. Because of this, terms such as "knock-down power" and "stopping power," which are frequently used in terminal ballistics, are actually meaningless. Knockdown power refers to the momentum transfer only, but in reality it is kinetic energy transfer that does the real damage. What happens to a target depends on the details of the collision and which of the above three cases applies, so you can't say a certain type of ammunition (or gun) has a certain knockdown power.

One of the major issues related to terminal ballistics is the penetration of the bullet. A measure of it is given by the impact depth of the bullet, which is the depth the bullet reaches before it is stopped. In some cases bullets are designed to achieve maximum penetration, in others they are designed to do maximum damage. Bullet design is quite different in the two cases. Bullets designed for maximum penetration are made so that they do not deform on impact (or at least, deform as little as possible). They are usually made of lead that is covered with a layer of copper, brass, or steel. The jacket usually covers only the front region of the bullet. In particular, armor-piercing bullets for small arms are usually made of copper jacketed with steel. For larger artillery such as tank guns, tungsten, aluminum, and magnesium are usually used in the shells.

Although some bullets are made to expand when they hit the target, this type of ammunition is now prohibited in warfare according to the Hague Convention of 1899, Declaration III.

HEY, LOOK . . . IT FLIES!

Aerodynamics and the First Airplanes

Soon after the first airplanes were invented they became important weapons of war. Early on they were used mostly for observation and reconnaissance, but it soon became obvious that they could play a much more important role. They could be used to release bombs on the enemy. It was a mere ten years after the Wright brothers flew their first airplane that the First World War began, but by then airplanes had already been used in warfare. In 1911 the Italians used an airplane to drop grenades on the Turks in Libya. And as it became obvious that airplanes would be useful in war, the technology associated with them advanced rapidly, and soon after World War I started they began to be used extensively by both sides.

DISCOVERIES THAT LED TO THE AIRPLANE

Although the most eventful day in airplane history was December 17, 1903, when the Wright brothers made their first flight in a power-driven, heavier-than-air machine, it was not the first attempt that humans had made to fly. Many important developments led up to that day, and I will begin with them.

We saw earlier the Leonardo da Vinci was obsessed with flight. Not only did he observe birds in flight for hundreds of hours, but he also studied the flow of both air and water around objects of many different shapes under various conditions. He noticed that water sped up as it moved around a rock in a stream, and he assumed that air did the same thing. Much of his effort went into trying to develop a pair of wings, like those of a bird, which a man could use to fly. He wasn't successful, but he did design a helicopter and a parachute, and both of these designs would have worked. Furthermore, he stated that the fluid dynamics are the same for an object moving through a fluid as they are for a fluid moving past the object in the same way. And finally, he also made an extensive study of drag, the frictional force an object experiences when moving through a fluid.

It was Galileo, however, who showed that the drag exerted on a body moving through a fluid is directly proportional to the density of the fluid, where density is mass per unit volume. The French scientist Edme Mariotte took this a step further in 1673 when he showed that drag is also proportional to the velocity of the object squared (v^2).

One of the most significant discoveries in relation to aeronautics, however, came in 1738 when Daniel Bernoulli of Holland showed that in a flowing fluid the pressure decreases as the velocity of the fluid increases. And of course this applies to all fluids, including air. It eventually became known as Bernoulli's principle. About the same time, the French chemist Henri Pitot demonstrated a device he called the pitot tube in which the change of velocity could easily be measured as the diameter of the tube changed.[1]

A further advance in the understanding of drag came in 1759 when the English engineer John Smeaton invented a device for measuring the drag produced on a paddlewheel moving through air. He showed that $D = ksv^2$, where D is drag, s is surface area, v is the velocity of the paddle, and k is a constant that became known as Smeaton's coefficient.

One of the most important people in the history of aeronautics, however, was the engineer George Cayley of England. He is usually considered to be the first person to understand most of the basic underlying principles and forces involved in flight, and because of this he has frequently been referred to as the father of aerodynamics. In particular, he discovered and identified the four major forces associated with flight: lift, weight, thrust, and drag. We will look at each of them in detail later. He also showed that "cambered" or curved wings produced the best lift. His three-part treatise titled "On Aerial Navigation," which was published in 1809 and 1810, was the most important early work on airplane flight. Most of the basic ideas associated with lift, drag, and thrust are discussed in it.

Although Cayley designed, made, and flew many gliders, it is Otto Lilienthal of Germany who is usually referred to as the "glider king." He made several important advances in hang gliders, and over his lifetime he made over two thousand flights in gliders of his own design. In August 1896, however, while making a flight, his glider stalled. He tried to regain control by adjusting the position of his body, but he failed. The glider fell to the earth from a height of fifty feet. He was conscious when help reached him, but he died soon thereafter.

The first American to make important contributions to aviation was Octave Chanute, a civil engineer from Chicago, Illinois. He published the book *Progress in Flying Machines* in 1894, which was the most complete survey of the research on heavier-than-air aviation up to that time. And although he designed many gliders and invented the "strut-wire" braced wing, he never flew any of his gliders himself. He is perhaps best remembered for the interest and encouragement he gave to the Wright brothers. Indeed, he visited their camp near Kitty Hawk, North Carolina, in 1901, 1902, and 1903—the critical years in the development of their first airplane.

THE WRIGHT BROTHERS

Although many men made important contributions, the Wright brothers of Dayton, Ohio, are credited with designing and building the first engine-powered heavier-than-air craft to successfully carry a man on an airborne flight. This occurred on December 17, 1903. Their major contribution to aeronautics is usually considered to be their invention of the three-axis control, which enabled the pilot to maintain equilibrium and to steer the aircraft effectively.[2]

Orville and Wilbur Wright spent their early years in Dayton, Ohio. They were the two youngest of eight children, and according to most biographers their interest in flying was sparked at a young age when their father bought them a toy helicopter that was powered by a rubber band. Wilbur was four years older than Orville. Neither man completed high school, but they became interested in newspaper publishing after they built a printing press. They started with the *West Side News* and later published other newspapers.[3]

Orville Wright.

In 1892 they opened a bicycle shop for the sale and repair of bicycles. A few years later, in 1896, they began selling bicycles that they had built. It was during this time that the work of Otto Lilienthal of Germany came to their attention. Lilienthal had built and tested several gliders. His work inspired them, and they began to read about Cayley's and Canute's exciting exploits in the field, and by 1899 and they had begun their own experimentation. As the older brother, Wilbur was the leader of the team.

Wilbur Wright.

Over the preceding decades several different approaches had been used in various attempts to fly, but the Wright brothers decided the best approach was to leave power-driven flight until they had solved all the problems associated with gliding. In particular, they believed that the pilot should have complete control of the plane at all times, using a system for banking, turning, and changing altitude. They decided to design such a system before adding an engine to the craft. Early on they discovered "wing warping," in which control lines were used to twist or warp the outer section of the wings so the plane could bank properly. Wing warping was controlled by four lines, set up so that the two wings work together. When the lift on one wing increased, the lift on the other wing would decrease.

When their glider was finally ready, they wrote to Canute, asking him where the best place was to test it. He suggested several places, but the one that interested them the most was Kitty Hawk, North Carolina. It had excellent breezes from the Atlantic that would be helpful, and it had soft sand to land on. They decided that it would be ideal, and in the autumn of 1900 they traveled to Kitty Hawk with their glider. It was a "double-decker" with two wings, and the top of each wing had a camber (curvature). It had no tail, since they saw little need for a tail at that time.

Both manned and unmanned tests were made, but when the glider was unmanned extra weight was added to account for a pilot. Wilbur was the pilot in the manned flights; he stretched out on his stomach across the lower wing. In all cases the glider was only about ten feet above the ground, and it had tether lines attached to it. They were particularly interested in testing the wing-warping apparatus that they had attached to the glider. As it turned out, they were extremely pleased at how well it worked. Having used Smeaton's equation for calculating lift, however, they were disappointed that the lift appeared to be much less than that predicted by the equation. Nevertheless, they were generally happy with their results but knew that improvements were needed.[4]

First glider of the Wright brothers.

Over the following months they worked feverishly to build a new glider. It had a much larger wingspan, and improvements had been made to the wing-warping apparatus. This time they arrived at Kitty Hawk in July, and during July and August they made about one hundred flights varying in distance from twenty to four hundred feet. Everything appeared to go well, but again they were disappointed with the lift the glider gave. It was well below—only about one-third—the value predicted by Smeaton's equation, and they began to wonder about the equation's accuracy.

One of the factors in the equation was a constant called Smeaton's constant. Its value had been worked out years earlier and had become the accepted value, but the Wright brothers were sure it was wrong, and there was only one way to prove it. They had to build a wind tunnel; and indeed, over the next year they built a wind tunnel in their bicycle shop. It was six feet long, and between October and December 1901 they tested two hundred different wing shapes, comparing their experimental results to the predictions of Smeaton's equation. And indeed they were right; Smeaton's constant was incorrect. Not only did they correct the equation, but they also learned a tremendous amount about wings. As Fred Howard, one of their biographers, said, "They were the most critical and fruitful aerodynamics experiments ever performed in so short a time with so few materials and at so little expense."[5] The tests also showed that longer, narrower wings were better than what they had been using.

With their new knowledge the Wright brothers eagerly began building a new glider. It had longer, narrower wings with a reduced camber. They also now realized that wing warping caused an additional drag at the wing tips, and they would have to take it into consideration. Finally, they also attached to their new model a tail with a rudder for steering. They soon discovered during the early tests, however, that the vertical rudder on the tail was important not just in turning, but also during banking turns and when leveling off after turns. They now had "three-axis control": wing warping for rollover, forward elevator flaps for pitch (up-and-down motion), and a rear rudder for yaw (side to side motion). And between September and October 1902 they made about one thousand tests. They were now ready to add an engine to the glider to power it.

What type of engine was best? It obviously had to be as light as possible, and because of this, they decided that it should be made of aluminum. They checked with several engine manufacturers, but none could produce the engine they wanted, so they decided to build it themselves. Fortunately, the shop mechanic in their bicycle shop was an expert on engines. They told him what they wanted, and within six weeks he had it completed. It was cast from aluminum, and as was common at the time, it had a very primitive fuel-injection system; the gasoline was gravity fed.

The Flyer I, as they called it, had a wingspan of just over forty feet and weighed 605 pounds, and it had a twelve-horsepower engine. It was built of spruce, with muslin covers over the wings. The propeller was eight feet long and had been designed for maximum lift. After some thought they decided to use a "pusher" design with the propellers mounted behind the pilot so that they pushed the craft rather than pulled it.

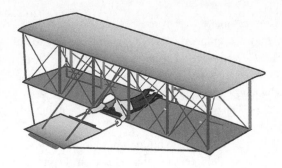

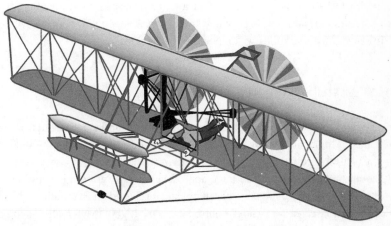

Top: 1901 model of Wright brothers' glider.
Bottom: 1903 model in which they made their first powered flight of twelve seconds.

When they were finally ready, they took the craft to Kitty Hawk, or more exactly, a part of Kitty Hawk called Kill Devil Hills, which consisted of sand dunes up to one hundred feet high. They arrived at their camp in early December 1903. After a delay created by a broken propeller, they began their tests on December 14. Again there were problems, but they were quickly overcome; then came the historic day: December 17. The first attempt was made by Orville; it consisted of a flight of 120 feet in a time of twelve seconds. The next two flights covered a distance of 175 and 200 feet respectively. Wilbur and Orville alternated as pilots. For the first time in the history of the world, man had flown a power-driven plane over a respectable distance. The Wright brothers were ecstatic; they telegraphed their father telling him to inform the press. Surprisingly, the *Dayton Journal* refused to publish the story, stating that the flights were too short to be important. But news of the event was leaked to other newspapers, which quickly published a highly inaccurate story to the dismay of the Wright brothers. The problems were soon ironed out, but strangely, the story created very little public interest at first.

The Wright brothers went on to build their Flyer II in 1904, and it was tested closer to home at an airfield about eight miles from Dayton. Without the "sea breezes" the takeoff was more difficult, so they built a weight-powered catapult to make takeoffs easier. The new airplane was more powerful, however, and they were soon flying it in circles. On September 20, 1904, they flew the first complete circle, but by December 1 they were covering almost three miles in four circles above the camp.

In 1905 they built Flyer III. It had a major improvement: all three of the axes—pitch, roll, and yaw—now had their own independent controls. A flight of almost twenty-five miles was made during these trials.

WHAT MAKES AN AIRPLANE FLY?

Most people have some sort of idea of how and why an airplane flies, but few really understand it in detail. There is, in fact, a lot of misinformation in books about this question. Lift can come from a propeller, a jet, or a rocket, but for now we will talk only about propellers. There are three approaches to answering this question. The first is what is called the simple explanation, and it is based on Bernoulli's principle, which we discussed earlier. And of course there's the highly mathematical approach based on various aeronautical principles—the approach aeronautical engineers use to design aircrafts—which is well beyond the scope of this book. The third approach is what we'll call the *physical explanation*, which is based on physics. It can get slightly complicated, but we'll try to keep it as simple as possible. In any case, it is the most accurate explanation at this level. It is based on the Bernoulli principle, but it shows that there's much more to it than just this principle.[6]

In this section I'll begin with the simplest approach because it's the easiest to understand. First of all, during the takeoff and landing of an airplane, and also when it is in flight, there are four forces acting on it: lift, weight, thrust, and drag. As the name implies, lift is a force that lifts the plane off the ground. It is created by the interaction between the wings and the air they pass through. According to Bernoulli's principle, the air pressure on the top of the wing decreases as the plane begins to move because the air traveling over the wing moves faster in comparison to the background air. The faster the plane moves, the greater the decrease in pressure upon the top of the wing. This creates a pressure difference between the top and bottom of the wing, which creates a net upward force.[7]

The four forces on an airplane.

Opposing lift is a force due to the weight of the airplane, namely gravity. During takeoff lift continues to build up as the upward force on the wing increases, and when it is finally greater than the weight of the airplane, liftoff occurs. In short, the plane leaves the ground.

The third force, namely thrust, is the force that moves the plane forward. It can be generated by a propeller, a jet engine, or a rocket, but for now we will confine our analysis to propeller-generated thrust. The propeller is curved in such a way that when it spins it pushes air backward. It is very much like the wing in that an air-pressure differential is created, with lower air pressure on the front of the blade and higher air pressure behind it. This creates a thrust that moves the plane forward. But, again, there is a force opposing it. As the plane begins to move through the air there is friction between it and the air, and this friction is called drag. Again, thrust has to be greater than drag for the plane to move forward, as the two forces oppose one another. As most people know, drag can be minimized by streamlining the shape of the moving object. A "teardrop" shape is among the best for minimizing drag.

If lift is greater than the weight of the airplane, and thrust is greater than drag, the airplane leaves the ground. So, basically an airplane flies because of the Bernoulli principle, which is the explanation given in most popular articles and books. But if you look carefully at this explanation, it's easy to see that it is not complete. Wings that have no camber (curve) also create lift, and if you calculate how much camber is needed for a small plane to lift off, you find that the distance over the top of the wing has to be 50 percent longer than the distance across the bottom. This looks like the wing shown in the diagram below, and we know that in most planes the distance across the top of the wing is only about 2 percent greater than the distance across the bottom.

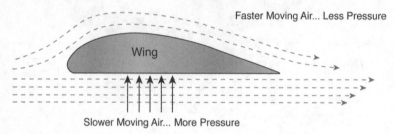

Faster Moving Air... Less Pressure

Wing

Slower Moving Air... More Pressure

The simple explanation of lift using the Bernoulli principle.

THE PHYSICAL, MORE EXACT EXPLANATION OF LIFT

The simple explanation above has many problems. First, it relies on what is called the principle of equal transit times. This states that the section or volume of air that goes over the top wing converges and joins at the trailing edge with the section or volume that goes under the bottom. Wind tunnel experiments show that this is not the case. The volume of air that goes over the wing reaches the end of the wing before the bottom volume reaches it.

Furthermore, the Bernoulli explanation ignores the fact that work is done by the lift. Lift obviously requires power and a force. And this relates to Newton's first law. As we saw earlier, Newton's first law states that a body at rest will remain at rest, and a body in motion will continue in a straight-line motion unless subjected to an external applied force. In the Bernoulli explanation there is no evidence of an external applied force. The streamlines above and below the wing are the same in this explanation. But in reality it's easy to see that there is a bend in the flow of the air; this means there is an acceleration, and therefore there must be a force acting on the wing (Newton's second law: $F = ma$).

Let's look at this force. Newton's third law tells us that for every action there is an equal opposite reaction. It's easy to show that the action in this case is what the wing does to the air. The reaction is the lift generated as a result of this action. We can understand this better if we go back to Newton's second law. The variables in our problem include the force on the wing, which equals the mass of the air moving downward times the change in air velocity. This is the lift on the wing, and it is effectively the amount of air moved downward per second times the downward velocity of the air. So, to a large degree, it's the downward velocity of the air that gives the lift. I should also point out that the downward velocity behind the wing is called the downwash (it creates an increased pressure), and there is an upwash at the front of the wing that also creates increased pressure.

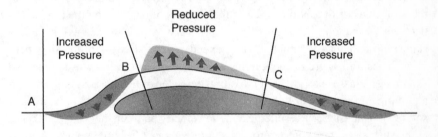

The physical, more exact explanation of lift,
showing positions of increased and reduced pressure.

Something else the Bernoulli explanation ignores is the angle of attack. This is the angle between the wing (or a line through the center of the wing) and the oncoming air. It has a large effect on lift. As the angle of attack increases, the air is deflected through a larger angle and the vertical component of the velocity is increased. This causes a large increase in lift. Eventually, however, at an angle of about 15 degrees it maxes out, and beyond this it decreases.

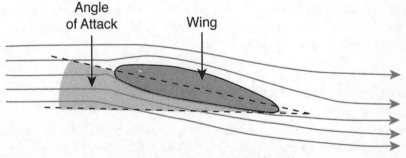

Diagram of wing showing angle of attack.

DETAILS OF DRAG

Drag is a mechanical force, and to be generated, the body must be in contact with the air. This is, of course, true in the case of the wing passing through air. Simply, it is friction between the air and the wing, and it is generated because of the difference in velocity between the wing and the air. Furthermore, it acts in a direction exactly opposite to the motion of the airplane. And finally, it is classified as an aerodynamic friction.[8]

Three types of drag friction exist: skin friction, form friction, and induced friction. Skin friction is the friction between the moving molecules of air and the molecules of the solid surface of the wing, so it depends on the interaction between these molecules. This means that a very smooth surface will have less skin friction than a rough one. It also depends on the viscosity of the air, where viscosity is a measure of the internal resistance of a fluid to deformation. For example, molasses has greater viscosity than water. When air is in contact with a moving surface, the air will try to follow the surface. In other words, it has a sort of "stickiness"; as a result, the relative velocity between the wing and the air at the surface of the wing is zero. As you move away from the wing, however, it gradually increases.

Form friction is an aerodynamic resistance to the motion of an object through air that depends on the shape of the object. The more streamlined the shape, the less the form friction. A teardrop shape has one of the lowest form frictions. This type of friction is particularly important in the case of cars; we stream them to decrease this type of drag as much as possible, which increases their fuel efficiency.

Induced friction occurs near the tip of wings that are curved or distorted. The "effective curving" causes a pressure difference between the top and the bottom of the region near the tip of the wing. It's called induced friction because it is induced by the action of vortices near the tip of the wings. Its magnitude depends on the shape of the wings and the amount of lift they produce. Longer, thinner wings produce less induced drag.

STEERING AND MANEUVERING THE AIRPLANE

As we saw earlier, after considerable experimentation the Wright brothers finally developed an effective three-axis control that allowed them to control and properly maneuver their airplanes. Wing warping was used for roll, or lateral motion, and a forward elevator on the wings was used to control up-and-down motion, or pitch, while a rear rudder was used to control side-to-side motion, or yaw. Within a few years, however, Glenn Curtiss of New York developed what are now called *ailerons* to replace the wing warping of the Wright brothers. Ailerons are small control surfaces that are attached to the trailing edge of the wing.

So how do we maintain control over an aircraft? As it turns out, takeoff, landing, and cruising each have to be stabilized and maneuvered differently. Wings are generally designed for an appropriate amount of lift, along with a minimum of drag, during cruising. But it's fairly obvious that things have to be quite different during takeoff and landing. The plane's speed is much less at this time, and the plane must be adjusted accordingly. And this is where flaps and slats come in. Without them the pilot would not be able to take off or land.

Flaps are hinged surfaces on the trailing edge of the wings that are used to reduce the speed of the airplane so that it can safely take off and land. They decrease the distance needed for both landing and takeoff. When they are extended downward from the trailing edge of the wing they effectively alter the shape of the wing so that it creates more lift on takeoff and more drag on landing.

Slats perform a similar function, but they are mounted on the front edge of the wing. Again, they are used to temporarily alter the shape of the wing to increase lift. In effect, they temporarily change the angle of attack of the wings. Using them, a pilot can fly at slower speeds during takeoff and can land in shorter distances.

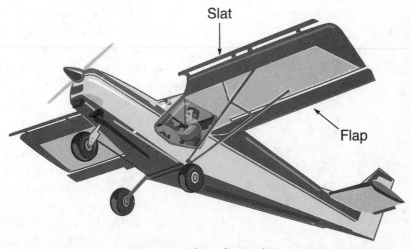

The slats and flaps of an airplane.

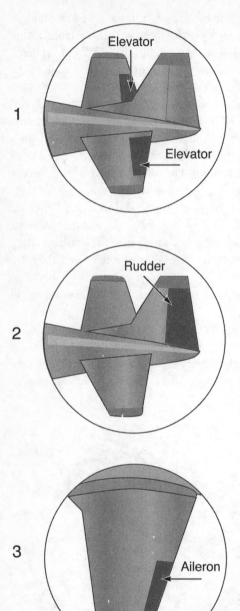

The tail, showing elevator and rudder, and the wing showing aileron.

The tail of an airplane has two small wings that are referred to as the horizontal and vertical stabilizers. They are flaps that are used to control the direction of the airplane. The flaps on the horizontal tail wing are called elevators, and they are used to move the plane in the up-and-down direction. They change the horizontal stabilizer's effective angle of attack, which creates a lift on the rear of the airplane, which moves the nose downward. The vertical tail wing performs the same function as a rudder on a boat. It moves the plane to the left and to the right.

Getting back to the main wings, we have the ailerons, which are located near the tail end of each wing. Again, they are flaps that allow the plane to bank, or lean, to the right or left. There is one on each wing, and they work in opposition to one another; in other words, when one is up the other is down. As a result, more lift is generated on one wing, which creates a rolling or banking motion.

An important question at this point is, how exactly does a plane roll? This also applies to other motions of the plane, and it brings us to the center of gravity of the airplane. As we saw earlier, this is the point within the plane where its entire weight can be considered to be concentrated. If you suspended the plane from this point

it would remain in equilibrium in any position. With
this in mind, imagine a line
from the plane's nose that
runs through the center of
gravity and out to the tail.
This is called the roll axis,
and when a plane rolls, or
banks, it moves around this
axis.

Now imagine another
line starting at the top of the
airplane that goes through
the center of gravity and
out through its belly. This
is called the yaw axis. It
is important when a pilot
moves the plane's rudder.
It creates a side force
that rotates the tail in one
direction and the nose in
the other. This motion
is around the yaw axis.
Finally, imagine a line that
is roughly parallel to the
wings and passes through
the center of gravity. It is
called the pitch axis. When
the elevators of the airplane
are changed (creating pitch)
the plane moves around the
pitch axis.

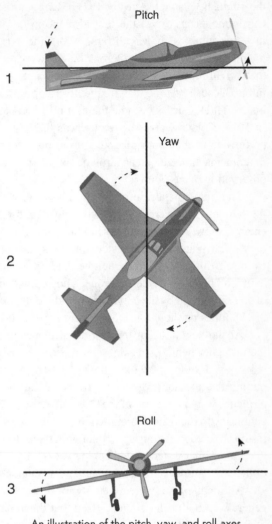

An illustration of the pitch, yaw, and roll axes.

FIRST USE OF AIRPLANES IN WAR

The most important aspect of airplanes, as far as this book is concerned, is their
use in war. Technology developed quickly after the early flights of the Wright
brothers. In 1909 the French pilot Louis Bleriot made the first flight across

the English Channel. And by 1911 Glenn Curtiss of New York had built the Curtis model D; it was the first airplane in the world that was built and sold in large numbers. It was different from the Wright brothers' airplane in one major way: Curtiss used ailerons on his wings instead of wing warping. Like the Wright brothers' model, however, it was a pusher with the wing behind the pilot. Because of this, it is frequently referred to as the Curtiss pusher. Using his model D, Curtiss became the first pilot to take off from and land on a ship. In 1913, Roland Garros of France was the first to fly over the Mediterranean; he flew from the south of France to Tunisia. In 1915, Garros was the first to mount a machine gun on an airplane. Within two weeks he downed four German observation planes with it.

The first official use of an airplane in war, however, came in 1911 when the Italians, who were at war with the Turks, used one to drop grenades on the enemy below. By the time World War I broke out in 1914, however, there were only a few planes on either side, and those available were relatively slow. The earliest British model, for example, had a top speed of seventy-two miles per hour, and it was powered by a ninety-horsepower engine. But technology developed rapidly, and by the end of the war the British SES, a fighter, had a top speed of 138 miles per hour and was powered by a two-hundred-horsepower engine.[9]

At the beginning of the war all airplanes were of the pusher design with the propeller behind the pilot, and most were used for observation and reconnaissance. It didn't take long, however, to discover that a propeller in the front, pulling the airplane, was more effective, and by the end of the war nearly all airplanes were pullers, or of the "tractor" design. All early planes also had rotary engines with the pistons mounted in a circle around the crankshaft. In effect, the pistons rotated, carrying a propeller with them. It was soon discovered, however, that the in-line, water-cooled engine was much more powerful. And by the end of the war, almost all airplanes had in-line engines.[10]

Although planes were used only for observation at the beginning of the war, generals on both sides soon realized that they could be used much more effectively, and they soon were used for tactical and strategic bombing, and for naval warfare. And, of course, the fighter plane was soon developed, and it played a critical role in the war.

13

THE MACHINE GUN WAR—
WORLD WAR I

In the last chapter we saw how airplanes were developed, and how physics is important in relation to them. In this chapter we will see how the airplane was used in war, and also in other ways. We now refer to the war that took place between 1914 and 1918 as World War I, but before 1939 it was known as the Great War, and it was, indeed, the largest, most all-encompassing war the world had ever seen. For years there had been a huge buildup of weapons in Europe, and with the new breakthroughs in science, many of the weapons had never been used in a large engagement before. Horrific scenes resulted, such as hundreds of soldiers being mowed down by machine guns within minutes, whole units wiped out by strange, new deadly gases, and ships sunk with new and powerful torpedoes. Perhaps the most ironic part of the war was its futility. Because of the new weapons—particularly the machine gun—the war soon became a stalemate, with neither side able to move. The opposing armies sat facing one another along hundreds of miles of trenches, with the trenches only a few hundred yards apart.

There's no doubt that many of the advances in weaponry were a result of breakthroughs in physics, but it was really the application of applied physics to weapons that led to the huge stalemate. Some of the weapons in which physics played a role were the machine gun, large cannons that now had become deadly accurate, airplanes, new types of rifles, hand grenades, flamethrowers, torpedoes, submarines, tanks, and new types of ships. And with these new weapons, over the four years of the war, millions of soldiers lost their lives without anything really being solved.

DEVELOPMENT OF THE MACHINE GUN

The machine gun played such a central role in the war that World War I is now sometimes referred to as the machine-gun war. The Gatling gun had been invented much earlier, but it was a toy compared to the 1870 invention of the British engineer Hiram Maxim. He devised a system that used the explosive gases set off by one bullet to propel the next one. The gun therefore ejected each spent cartridge and inserted a new one. Ammunition was fed into the firing chamber using a belt-fed system. A water-filled steel jacket surrounded the barrel; it kept the temperature cool enough so the barrel would not crack or melt as a result of the intense heat generated by the explosive gases.[1]

The major problem with the Maxim gun was that it was relatively heavy and difficult to use. In particular, it needed several men to operate it, and it wasn't highly reliable. Because of this, it had seen limited use before the beginning of World War I. In 1896, however, the Maxim was improved quite significantly. The Vickers Company in England purchased the Maxim Company, and Vickers redesigned and improved the gun. First, Vickers decreased the gun's weight by using lighter metals in its construction, and then he simplified the action. Basically it was now a water-cooled gun that used .303 British shells, the same shells that were used in the standard British rifle, the Lee-Enfield.

Although it took six men to operate it—one to fire it, one to feed ammunition, and four to move it around and set it up—it was quite reliable once it was set up, and as a result it became a favorite weapon among British troops. It was three feet, eight inches long with a firing rate of 450 to 600 rounds per minute, and it had maximum range of about 4,500 yards. It could be fired for twelve hours without overheating or breaking down, and about ten thousand rounds could be fired each hour. At the end of this time, however, the barrel had to be changed. It was particularly effective against troops in the open, and it was one of the major reasons for the stalemate in the war. Later in the war (after 1916) it was used in airplanes, both by the British and the French.[2]

Another machine gun referred to as the Lewis gun was also used extensively. It was an American-designed gun that was much lighter than the Vickers. It was widely used by the British. Strangely, although it was designed by a US colonel named Isaac Lewis in 1911, it was not used extensively by American troops in the war. Lewis, in fact, became disillusioned with the US Army in 1913 because its leaders refused to adopt his gun, and he left the United States, went briefly to Belgium, and then went to the United Kingdom, where he worked with British manufacturers to build the gun. At twenty-eight pounds, it was only about half the weight of the Vickers, and its length was barely over four feet, so it could easily be carried by a single soldier. It used the British .303 bullet (although some models use the .30-06), and its rate of fire was about six hundred rounds per minute, with an effective range of about 880 yards and a maximum range of 3,500 yards.

The French 75 millimeter field gun also played an important role in several battles, particularly near the beginning of the war. It had a recoil mechanism that allowed the barrel to slide back and forth after it was fired before returning to its original firing position. Because it did this without moving the gun, no re-aiming was required. It was tremendously accurate and could fire fifteen shots per minute. For anyone in the open it meant certain death, and in the Battle of Marne over two thousand German soldiers were mowed down in four minutes.

OTHER WEAPONS

It may sound like the machine gun was the only weapon used in World War I. This wasn't the case, and some of the other guns were almost as lethal as the machine gun. Rifles had improved significantly, along with cannons and other artillery, and hand grenades began to play a large role. Flamethrowers were also used. And, as mentioned earlier, it was the large array of new and powerful weapons that led to the stalemate. Furthermore, most of them were a result of breakthroughs in science—particularly physics.[3]

The muskets of the Civil War were replaced on the British side by the bolt-action Lee-Enfield rifle. The name comes from its inventor, James Perris Lee, and the factory in Enfield, England, where it was first produced. Cartridges were placed in a metal box called the magazine, with a spring at the bottom. When the bolt was open, the spring pushed the cartridge upward into position. As the bolt was closed, the top cartridge was pushed into the chamber, ready to fire. After firing, the bolt was opened, causing the empty cartridge to eject, and a new cartridge was loaded. The magazine held ten shells and was loaded with .303 British ammunition.

The bolt-action was fast and easy to use, and with its relatively large detachable magazine, it was an excellent rifle. A well-trained rifleman could easily fire twenty to thirty rounds a minute with it. It was accurate up to two thousand feet, and it had a range of about 4,500 feet. The Germans used the Mauser Gewehr, which was also a bolt-action gun that was well known for its accuracy. And they also used the Mauser T as an antitank gun toward the end of the war. In addition, pistols such as the British Webley, the German Lugar, and the American Colt .45 were used extensively during the war. They were usually carried by officers.

Cannons and large artillery had also been improved significantly. Howitzers, which are cannons that fire shells in a high, curving trajectory, were used quite effectively against fortifications and hidden targets. They fired heavy shells through relatively short barrels, and were used by the Germans early in the war in Belgium.

The huge howitzer known as "Big Bertha" pounded fortresses in Belgium during the German advance through the country. The biggest problem with the large guns was their weight, but this was finally overcome by using railroads. Many were, in fact, mounted on railroad cars, which also helped to overcome their recoil. Such guns could fire huge shells to distances of up to thirty miles.

The mortar was also used during World War I. Like the howitzer, it was a high-trajectory gun, but much smaller. The projectile could easily be dropped into its broad, stubby barrel and fired quickly. It could fire shells as far as twenty-two hundred yards. And antiaircraft guns were also developed soon after airplanes started to play a large role in the war. They could fire four rounds per minute to a distance of three thousand yards.

Crude hand grenades had been around for years. The early Chinese used them, and they were also used by the French in the fifteenth century. Considerable effort went into perfecting them during World War I. The Germans were well prepared with hand grenades at the beginning of the war, but other countries quickly caught up. Trench warfare, in particular, made them very effective. "Bombing parties" from both sides would attack the enemy trenches using hand grenades of various types, and these missions increased in frequency as the war progressed. The British had very few hand grenades at the beginning of the war, but within a year they were producing half a million a week.

Grenades could be detonated in either of two ways: on impact (percussion) or using a timed fuse. The fuse-type grenade was generally preferred because of the danger of setting off the percussion-type grenade accidentally. In the timed grenades a pin that could be extracted by hand became common in the later stages of the war. And they came in all shapes and sizes. The stick grenade had a handle, one of the others was cylindrical, but later on most were oval shaped. Grenades could be either thrown or launched with a rifle. Rifle grenades were attached to a rod that was placed down the barrel of a gun or placed in a cup that was attached to the barrel. A blank cartridge was used to launch them. Cup grenades were particularly popular with the British and French; although they were not very accurate, they could be thrown a couple hundred yards by the blast.

The Mills grenade was introduced by the British in May 1915 and quickly became very popular. It had a serrated exterior so that when it detonated it broke into many fragments, and it was fairly light at 1.25 pounds. A safety pin held down a strike lever. After the pin was removed, the strike lever was held down by hand until the grenade was thrown. It had a four second fuse.

The Germans also had an array of different grenades: stick, ball, disk, and egg shaped. They preferred the egg-shaped grenade because it could relatively easily be thrown up to fifty yards.

Perhaps the greatest terror of the war, at least for the soldiers it was used against, was the flamethrower. Although crude flamethrowers had been used in earlier wars, this was the first war in which an efficient, well-designed flame-thrower was used. It was used by the Germans against the British and French soldiers in the early phases of the war in 1914. The Germans had begun experi-menting with flamethrowers as early as 1900. They used pressurized air, carbon dioxide, or nitrogen to push oil through a nozzle. As the mixture hit the air it was ignited by a small trigger and became a jet of flame. Early flamethrowers had a range of about eighty feet, but this was later increased to about 130 feet. This made them quite effective in trench warfare.

The Germans had two models, a relatively small portable one that could be carried by a single man, and a larger, much heavier one, that had an effec-tive range double that of the small one. It required several people to transport it. The first use of the smaller variety came at The Hague in Flanders, Belgium. It occurred on July 30, 1915, when Germans with gas cylinders strapped to their backs attacked a British line. The huge flames terrified the unsuspecting British at first, but, after losing some ground, they managed to hold their position. The Germans were pleased with its success and started using it in most of their suc-cessive battles. The men armed with flamethrowers soon became marked men, however; British and French soldiers concentrated their fire on the operators, and few lived very long.

The British soon began experimenting with their own flamethrowers. They constructed several models, ranging from a relatively light portable model up to very heavy ones. The larger ones had a range of about ninety yards. The French also developed several models. They were quite effective in the Battle of the Somme in France.

Other weapons that were used extensively included poisonous gas and tanks. They will be discussed in a later section.

HOW THE WAR STARTED

The spark that ignited the war was the assassination of Archduke Franz Ferdinand, heir to the Austro-Hungarian throne, at Sarajevo on June 28, 1914. It led to a sequence of almost mindless events that occurred very quickly, mostly because all the nations involved had treaties and alliances with other nations. Even though Franz Ferdinand was quite unpopular, the Austro-Hungarians immediately accused the Serbians of a conspiracy (Ferdinand was actually killed by a young terrorist from a group called the Black Hand) and issued several ultimatums to them. The Serbs rejected some of the ultimatums, and, as a result, the Austro-Hungarians mobilized their army on July 28, 1914. But Russia, bound by a treaty to Serbia, quickly came to the rescue. In the same way, Germany had a treaty with Austria-Hungry, and, as a result, Germany was likewise drawn into the war on August 1. Then came France, which had a treaty with Russia; it declared war on August 3. As a result of Germany's invasion of Belgium, Great Britain was pulled into the war when the King of Belgium appealed to Britain for help.[4]

Everyone thought at first that it would be a relatively brief war, but it soon escalated into an uncontrollable nightmare, mostly because of the stubbornness of the nations involved. The three major continental powers, Germany, France, and Russia, all went on the offensive immediately. But to their surprise, they discovered that the new array of lethal weapons that each country now had gave no advantage. Machine guns began mowing down troops by the hundreds. The best defense proved to be a deep trench. Within a short time the front had become two lines of trenches a few hundred yards apart, with neither side wanting to attack. And strangely, for the next four years these trenches moved very little. It was a stalemate. Indeed, in the few attempts at moving the line, hundreds of thousands of troops were slaughtered. Not only were they mowed down by machine guns and other artillery, they were also gassed and bombarded by hand grenades and flamethrowers. And overhead there was something new: airplanes soon began strafing the trenches with bullets.

THE FIRST WARPLANES

The first powered aircraft had been flown by the Wright brothers only a decade before World War I began, but the airplane was soon to play a major role. At first planes were only used for observation and reconnaissance, and indeed they were able to provide an important new view of the battlefield. At the Battle of Mons in southern Belgium on August 23, 1914, British forces rushed to the rescue of the French as the Germans attacked them. Just before the clash, the British sent out an observation plane to see what was going on and discovered, to their surprise, that the Germans were trying to surround them. The British high command immediately ordered a retreat, which saved them from a disaster. A little later a French observation plane noticed that the Germans had exposed their flanks, and the French attacked, stopping an attempted drive to Paris. The value of the observation plane was soon evident.[5]

It wasn't long, however, before observation planes on opposite sides began to come in contact. At first they merely fired at one another with pistols and rifles, although at times they also tried to throw rocks at one another's propellers. One of the first pilots to escalate air warfare was Roland Garros of France. Although most of the early airplanes used in the war were "pushers," like the Wright brothers' craft, with the propeller behind the pilot, it was soon discovered that the "tractor" design, with the propeller in the front, was much more effective. The problem with this, however, was that if machines gun were to be mounted so pilots could easily aim and fire them, they would have to fire through the whirling blades of the propeller, and this would quickly destroy the propeller blades. Garros decided to protect the blades by adding steel deflectors to them.

In early April 1915 he tried out his new invention for the first time. The recipients of his attacks were no doubt surprised when they saw Garros's airplane flying straight at them, shooting a stream of bullets. Garros shot down four German airplanes using his new device, but on April 18 he was forced down behind German lines. His airplane was seized, and the German high command called in Anthony Fokker of the Fokker aircraft company and ordered him to copy the device. Fokker saw, however, that it was seriously flawed: many of the bullets hitting the blades were deflected, and some of them were being deflected backward. Fokker and his team therefore began working on a system in which the machine gun was synchronized with the propeller blade. A cam was placed on the crankshaft of the airplane; when the propeller was in a position where it might be hit by a bullet, the cam actuated a pushrod that stopped the gun from firing. The new device was placed on German airplanes, and for many months the Germans had a tremendous advantage in the air.[6]

In the meantime the British were also experimenting with the mounting of a machine gun on an airplane. Aviator Louis Strange attached a machine gun to the top of the upper wing of his plane so that the bullets would clear the propeller. However, on May 10, 1915, his gun jammed. He stood up on his seat in an attempt to pry it loose, but as he worked on it the plane suddenly stalled and flipped over, then it began to spin downward. Strange was flung out of the plane, but he managed to hang on to the gun on the upper wing. For several moments he swung his legs around wildly, trying to get back in the cockpit. Fortunately, he succeeded and was able to pull the plane out of its dive just before it would have crashed.

Nevertheless, with the new Fokker design, the Germans quickly achieved air superiority. Strangely, most of their new guns were mounted on the Eindecker G, a plane that was generally inferior to most British aircrafts. The casualties, however, were not as great as they were later in the war because the British pilots stayed clear of the Eindecker. The morale of the British, however, had been shaken, and they rushed to produce fighters that could match the Eindeckers.

The era of the dogfight had begun. A dogfight was an aerial battle between two or more aircraft. The Germans had an advantage when the dogfights first began. As a result, they began knocking down British planes at the rate of the about five to each of their losses. German aces such as Max Immelmann and Oswald Boelcke became heroes at home as a result of the large number of British planes they downed. Between them they shot down almost sixty enemy aircraft before they were stopped. Finally, though, in the fall of 1915, the British introduced two fighters, the FE 26 and the DH2, which were a good match for the German planes, and they also developed tracer ammunition that helped. With it a pilot could see his stream of fire and adjust it if needed.

Pilots with eight kills became known as aces.[7] At first, most pilots went out alone, searching for enemy planes, but after 1917 squadrons were introduced on both sides. The British developed squadrons of six planes that usually flew in a V-shaped formation with a commander in the front. In combat, however, they would break up into pairs, with one of the two planes primarily on attack while the other was a defender. The German squadrons were usually larger, and their groups eventually became known as circuses.

One of the leading British aces was Mick Mannock. He was a leading developer of British air tactics, and between May 1917 and July 1918, he shot down seventy-three German planes. Almost all pilots were under the age of twenty-five, with many as young as eighteen. Many, in fact, were sent into battle with as little as thirty hours of air training. So, as expected, their life expectancy once they joined up was not long.

Dogfight tactics were well known, and everyone used them as much as possible. The major tactic was diving toward another plane from above when the sun was shining in the eyes of the opposing pilot. Both sides also used clouds for cover as much as possible; they would attack, then head for the clouds.

The best-known ace of the war was, no doubt, the German Manfred von Richthofen, who was also known as the Red Baron. He was credited with eighty combat victories during his career. As in the case of most aces, however, many of them were against greenhorn pilots with only a few hours of experience. He did, however, down one of the leading British aces, Major Lance Hayden. During 1917 he was the leader of the German squadron called the Flying Circus. The plane he piloted was painted red, both inside and out. His career came to an end on April 21, 1913, when he was shot down by ground fire near the Somme River.

On the Allied side, Billy Bishop was one of the most celebrated aces; a Canadian, he was credited with seventy-two victories, and he was instrumental in setting up the British air-training program. At one point he fought against the Red Baron, but neither man gained a victory. He was awarded the Victoria

Cross in 1917. The best-known American ace was Eddie Rickenbacker. Before he became a pilot he was a racecar driver, so flying a fighter plane was a natural next step for him. When the United States entered the war in 1917 Rickenbacker enlisted immediately and was soon flying over Germany. On September 24, 1918, he was named commander of a squadron. In total, he shot down twenty-six German aircraft. Another major American figure was Billy Mitchell; by the end of World War I he commanded all American air combat units.

Although the fighter planes got the most attention during World War I, a much larger plane also played an important role. It was developed to carry and release bombs over enemy territory. Strategic bombing was used quite extensively during the war. Its object was to destroy factories, power stations, dockyards, large installations of guns, and troop-supply lines. The first bombing missions were by the Germans, who launched terror raids using large Zeppelins (huge balloons) to bomb small villages and civilians as a way to destroy the enemy's morale. There were a total of twenty-three of these raids over England, and at first there was little defense against them. But quite quickly it became evident that they were easy targets, as most were filled with flammable hydrogen and therefore easy to shoot down. So the airship raids finally stopped, but Germany soon developed bomber airplanes. The British also developed the Handley Page bomber in 1916, and in November of that year they bombed several German installations and submarine bases. By 1918 they were using four-engine bombers to attack industrial zones, with some of the bombs weighing as much as 1,650 pounds. They developed a squadron that was able to penetrate deep into Germany and hit important industrial targets. The Germans retaliated, bombing both British and French cities, but in the end the British dropped 660 tons of bombs on Germany—more than twice the amount the Germans dropped on England.

THE WAR AT SEA AND THE MENACE BENEATH THE SEA

When the war began the British immediately set up a blockade in an attempt to stop materials and supplies from getting to Germany. And it worked fairly well, particularly in the later stages of the war. The British navy was given the job of enforcing the blockade; it was, after all, the most powerful navy in the world, and for many years it had been the model for other countries. Furthermore, at the beginning of the war its battleships and cruisers easily outnumbered those of Germany: twenty-one large battleships and nine cruisers to Germany's thirteen battleships and seven cruisers. Germany had no interest in meeting the British navy head-on at sea, and so it generally kept a low profile. Nevertheless, both sides expected considerable conflict, but an incident that occurred at the time gave some indication of what would come.

The British navy trapped the German battle cruiser SMS *Goeben* and a light cruiser in the Mediterranean. British officers knew that it was likely going to try to break through near Gibraltar, and they waited for it with their guns ready. Although the *Goeben* had slightly larger guns and greater speed than the British ships, it was outnumbered, and it looked like it was easy prey. But to the surprise of the British, it held them at bay with its superior range, and it easily steamed around them unharmed. In fact, on its way out it picked off a couple of the British ships. The British were stunned, and it was soon obvious that they were no longer lords of the sea. They hadn't fought a battle in a hundred years and were obviously not well prepared for war.

So there were no great sea battles between German and British battleships, but for the next few years there were problems for the British navy. By 1914 the Germans had the best submarines in the world, and although the British didn't take them seriously at first, they were a formidable enemy. In fact, they became a real thorn in the side of the navy, and British leaders weren't quite sure what to do about them. Furthermore, it soon became obvious that they had almost no defense against them. On September 22, 1914, three large British armed cruisers were sunk in less than an hour by the German submarine *U-9*. Fourteen hundred men drowned, making it one of the worst disasters for the British navy in three hundred years.

Large numbers of German U-boats were now being seen in the North Sea, but they still weren't taken too seriously. For the most part, the submarines were warning their targets that they were going to be sunk, so the loss of life was relatively low. But in November 1914, the German chief of naval operations decided to do away with the warning, and in February 1915, all waters around Great Britain and Ireland were declared a war zone. This meant that all merchant vessels, including neutrals, would be sunk without warning, and over the next four months the Germans sank thirty-nine British ships. This became a serious problem for the British, and they soon began to look for better defensive measures. Then came the sinking of the luxury liner *Lusitania*, and they knew they had to do something fast.[8]

On May 1, 1915, the *Lusitania* began its 202nd crossing of the Atlantic. It set out from New York and headed for Liverpool, England, with 1,257 passengers and a crew of 702. They would never reach their destination, and most would perish. Waiting in the waters near Ireland was the German submarine *U-20*. On May 7, when the *Lusitania* approached the Irish coast, the captain of

U-20 spotted it coming toward him, and as it approached it turned thirty degrees, making it an easy target. Within seconds a torpedo was on its way. Several people on the *Lusitania* saw bubbles in the water as the torpedo approached, and someone yelled, "torpedoes on the starboard side." A large explosion followed, then a second muffled explosion came from the bottom of the ship. Almost immediately the ship tilted by twenty-five degrees, making it difficult to launch lifeboats. One thousand one hundred and ninety-eight people perished—almost as many as in the sinking of the *Titanic*.[9]

America was particularly enraged because 128 of those lost were Americans. President Woodrow Wilson sent a strong protest to Germany, threatening to break off relations if there wasn't an immediate stop to the German attacks on neutral ships, and the British were of course outraged. Other nations added their condemnation, and much to the surprise of England, the Germans called off all U-boat activity, and for almost a year and a half no British ships were sunk. This gave the British time to develop effective weapons against submarines. By June 1917, they had developed hydrophones that allowed them to pick up the sound of the propeller of a submarine from beneath the surface. Then they developed depth charges, which were oil drums filled with TNT set to explode at certain depths. In addition, they developed a device for throwing numerous depth charges outward from the deck of the ship.

The British were therefore ready when Germany resumed submarine warfare in January 1917. Less than three months later the United States entered the war, and by then another technique had also been developed to protect ships from submarines: the use of convoys with destroyer escorts. A submarine was lucky if it could sink a single ship of the convoy, and when it tried, it was in constant danger from depth charges. In addition, large numbers of mines were laid at depths up to six hundred feet in the North Sea and in the region between Scotland and Norway. U-boats soon became completely ineffective.

THE FINAL HORROR—POISONOUS GAS

The stalemate was a serious problem for both sides; each wanted to attack but knew that it would be suicide unless some sort of new weapon was developed. In their frustration the German high command turned to the physical chemist Fritz Haber. He had helped them earlier with a problem related to their ammunition, and German leaders hoped he could help again. Was there something that could be fired into the Allied trenches that would force the troops in them to flee? Haber immediately thought of poisonous gas. Several German generals had reservations, saying that the Allies would no doubt retaliate the same way, but Haber assured them that their chemical industry would have a hard time producing a similar gas. Despite these reservations, Haber was told to produce the gas. Haber decided on chlorine, and the Germans introduced it in April 1915, near Ypres. This region of the line was held by a combination of British, Canadian, French, and Algerian troops. Thousands of tanks containing chlorine were transported to the German lines. Fans were then used to blow the poisonous gas toward the enemy.[10]

On the evening of April 22, 1915, French and Algerian troops noticed a large yellow-green cloud drifting slowly toward them. Puzzled, they became suspicious that it was being used to conceal offensive troop movements, and they stood their ground waiting for an attack. Within minutes the cloud was all around them, and they were choking and gasping for breath. The inhaled gas was destroying their respiratory organs. When they realized what was happening they began to panic, and many of them fled in disorder. Within a short time a four-mile gap was opened along the line. Strangely, though, the Germans were as surprised by the effectiveness of the gas as the French were. And although German troops did advance, they did so nervously and hesitatingly. They managed to seize some land, but the British and Canadians on the right fought valiantly, and in the end little was gained. But a new phase in the war had begun.[11]

The British press immediately condemned the attack and played up the incident, and other countries, including the United States, soon joined in. Despite the condemnation, though, the British immediately went to work on research into poison gases that they could use to retaliate. There was, however, a problem with the delivery of the gas. If the wind changed direction while it was being delivered, it could blow back upon the troops delivering it. And indeed this happened to both the British and German troops. A better delivery system was needed. Again, the German high command went back to Haber. Was there a poisonous gas that could be easily packed into artillery shells and exploded in the trenches? Haber and his team went to work immediately and soon came up with phosgene. It was similar to chlorine gas, but unlike chlorine it caused no coughing or choking while it destroyed the lungs. As a result, soldiers usually inhaled much more of it before they realized what was happening. As a weapon it was therefore much more potent.

Then Haber came up with the most dreaded gas of the war—mustard gas. The Germans used it for the first time against the Russians in September 1917. Mustard gas was almost odorless, and it caused serious blisters both internally and externally.

Each time the Germans developed a new poisonous gas, however, the Allies soon developed the same gas and used it on the Germans, so in the end it was of little advantage to either side. The Germans inflicted several hundred thousand casualties, but they suffered around two hundred thousand themselves, with about nine thousand deaths.[12] Toward the end of the war, however, there were few casualties, as gas masks had been developed.

Although Haber apparently never felt any guilt for his role in developing poisonous gas, his wife was so appalled at what he had done that she committed suicide. His close friend Albert Einstein also severely reprimanded him for his role in slaughtering so many fellow human beings. But in the end it backfired for him. He was of Jewish descent, and in 1933 he had to flee Germany as the Nazi's began rounding up Jews.

THE FIRST TANKS

Another attempt at breaking the stalemate came with the introduction of the first tanks to the battlefield in late 1916. The idea that an armored, bulletproof vehicle could be helpful in battle had been around for many years. Even Leonardo da Vinci designed one. But it was not until after World War I had started that the idea began to be taken seriously. The impetus came from a British officer, Colonel Ernest Swinton. While driving through northern France in October 1914, having seen the large number of casualties inflicted by modern weaponry, he began to think about how troops could be better protected. A friend had mentioned a vehicle he had seen with large caterpillar tracks, and he suddenly realized that it would be extremely helpful to build a bulletproof military vehicle with caterpillar tracks.[13]

In November he talked to Lieutenant Colonel Maurice Hanley about the idea; Hanley, in turn, sent a memo to the Committee of Imperial Defense. The army, however, had little interest in the new device. Swinton therefore organized a demonstration of a caterpillar-track device for several high-ranking dignitaries in June 1915. In attendance were Lloyd George (minister of munitions, who would eventually become prime minister) and Winston Churchill, the first lord of the admiralty. Both were impressed, and Churchill immediately established what he called the Landships committee to look into the building of such a device. It didn't take long for the committee to decide the new device could be helpful in the war effort, and they agreed to go ahead with the design and building of a prototype model. It was important to keep it secret, so they code-named it "tank," so the Germans would not know what it was. And of course, the name stuck.[14]

Swinton was hired as an advisor, and he suggested several criteria for the new vehicle. It had to have a minimum speed of four miles per hour; it had to be able to cross a four-foot trench and pass easily through barbed wire; and it had to be capable of climbing over objects five feet high. In addition, it had to be bulletproof, and it had to have two machine guns. When it was finally built it was nicknamed "Little Willy." It didn't quite live up to Swinton's criteria, but it was close. It could move at a rate of about three miles per hour on level ground; it weighed fourteen tons and had a twelve-foot-long track frame; and it was rhomboid shaped and could carry three people. In early tests, however, it had problems crossing trenches, but this was soon corrected in a slightly larger model called "Big Willie." Of particular interest was that these tanks were produced by the navy rather than by the army.

Everything was now ready for the production of combat models, and the first combat model, called the Mark I, was demonstrated in June 1916. Lloyd George was impressed enough to order the immediate production of full-size tanks. In the meantime, the French had heard of the English plans and had begun work on their own tank.

The Mark I was ready for battle in September 1916, with thirty-six having been completed. Several military leaders, including Churchill, urged complete and thorough testing before it was used, but others wanted to use it as soon as possible. At the time the battle of Somme was in progress in France, and it hadn't gone as well as the British had hoped. As a result there was pressure to begin deploying the tanks. Thirty-six tanks therefore lined up on the front lines at Flers in France in September, and their appearance no doubt stunned the Germans. But Churchill was right: they were not ready. Many of them broke down in the initial attack, and some of them got stuck in the mud, so aside from the shock value, they were not very effective.

Meanwhile the French, having produced 128 tanks by April 1917, took them into battle, but as in the case of the English tanks, they were also not ready, and several problems developed. The first really successful use of tanks came on November 20, 1917, at the Battle of Cambria. With a force of 474 tanks the British attacked the German lines and breached a twelve-mile stretch. In the process they captured ten thousand German soldiers and a large number of machine guns. Surprisingly, though, the Germans did not try to emulate their attackers; they were slow in coming up with their own plans for a tank. One of the reasons, no doubt, was a lack of resources at this stage of the war.

The British and French, on the other hand, poured all their resources into the production of tanks. By the end of the war, the British had built 2,636 and the French had built 3,870. In addition, the United States built eighty-four tanks.

The Germans, on the other hand, only produced twenty, but they did develop weapons that were fairly effective against it.

All in all, the British Mark I performed fairly well considering the conditions. Most battlefields were littered with huge craters and strewn with barbed wire. The Mark I was able to move quite effectively over the very rough terrain, and it could easily cross trenches and craters of up to nine feet, and it had no problems with barbed wire. Indeed, it could even knock down small trees.

AMERICA ENTERS THE WAR

The beginning of the end began at the Eastern front, where the Russians had been fighting the Germans for two and a half years. They had suffered several defeats and morale was at an all-time low. The Russian army was in shambles, and the government back home was falling apart. As a result, in March 1917, Czar Nicholas was removed from power and a provisional government was set up. Surprisingly, though, despite the problems and setbacks the war had created, the new government vowed to fight on. But by now the Russian army was starting to fall apart; desertions were becoming more and more common, until finally the generals decided that they had had enough. Seeking peace, the Russians signed the Treaty of Brest-Litovsk.[15]

Germany was now able to send large numbers of troops to the Western front. They began arriving at the rate of ten divisions a month, and with the new build up, German high officials decided it was time to hit the Allies with a decisive blow so overwhelming that they would quickly agree to end the war. And on March 21, 1918, they struck. Within days a huge gap was opened up between the British and French lines. The Germans pushed forward, trying to take advantage of it, but they were surprised by the resolve and stubbornness of the British troops. They held their line. Meanwhile, the United States had declared war on Germany, and American troops were now coming across the Atlantic in large numbers.

For years President Woodrow Wilson had argued that the United States should stay out of the war, and most Americans agreed with him. But after the sinking of the *Lusitania* (with 128 Americans aboard), many Americans were angered, but Germany quickly stopped U-boat attacks, causing the anger to subside. On January 31, 1917, however, Germany decided to restart their unrestricted war on all shipping vessels in the war zone, neutral or not. President Wilson was stunned by the news, but the United States refrained from declaring war. But in February and March German submarines sank several American ships. In addition, British intelligence had intercepted and decrypted a message from Germany to the Mexican government. The Germans were promising Mexico that in return for its support all territories it had lost to annexation by the United States would be returned. These territories included Texas, New Mexico, and Arizona.

On April 2, 1917, Wilson asked Congress for a declaration of war, and it was quickly granted. The first American troops began crossing the Atlantic a few months later. A contingent, commanded by General Pershing, landed in France in June. The Germans were still attacking the Allied line, but now with considerably less success. And soon they were facing the first American troops.

A uniform Western command was formed in April 1918 that included British, US, and Belgian troops under General Foch. In the meantime the number of Americans in France doubled in March, and then it doubled again in May and August. The Germans attacked again in July but were quickly pushed back by a counteroffensive. The British then advanced in the north, and the Americans went on the offensive throughout the Argonne region of France. On July 18 Foch's forces along with the French went on the offensive along with nine double-strength American divisions. The Germans began to weaken, and then on August 8, over four hundred British tanks faced them. Soon they were surrendering by the thousands. And Germany's allies began to surrender: first there was Belgium in September, then Turkey on October 30, and finally Austria-Hungry on November 4. German morale plummeted as its resources also began to collapse, and finally, on November 11, 1918, the Treaty of Versailles was signed, ending the war.

THE INVISIBLE RAYS

The Development and Use of Radio and Radar in War

Electromagnetic radiation has played a large role in warfare since World War I, especially as used in radar, radio, and lasers. To understand these technologies, however, we have to go back several years before the beginning of World War I.

THE PRODUCTION AND DETECTION OF ELECTROMAGNETIC WAVES

James Clerk Maxwell is regarded by many as one of the most important physicists ever born. His prediction of the existence of electromagnetic waves led to major advances in science and also to important changes in everyday life.[1]

In the mid-1800s, four basic things were known about electricity and magnetism:

- All electric charges are surrounded by an electric field. The direction of the field is such that like charges repel and unlike charges attract.
- Magnetic poles of two types exist, referred to as north and south, and they always exist together.
- A changing electric field (or a charge) generates a magnetic field.
- A changing magnetic field generates an electric field.

These facts were known before Maxwell's time. His contribution was to put them in mathematical form and show that electricity and magnetism are intimately related, together forming what we call an electromagnetic field. In particular, oscillating electric charges produce an electromagnetic field that moves out from the charges, and the waves produced have both an electric and a magnetic field associated with them. Of particular importance, Maxwell found that electromagnetic waves traveled at the speed of light, and he proposed that light itself was an electromagnetic wave. Furthermore, he pointed out that electromagnetic waves of higher and lower frequencies (frequency at which the charge was oscillating) would likely lie beyond the frequency of light. In other words, there should be an array of electromagnetic waves of all different frequencies. And indeed, we now know that this is the case.

His prediction was made in the 1860s, and it wasn't long before electromagnetic waves were detected directly. In August 1879 the German physicist Heinrich Hertz built a simple device in his lab that he believed could be used to detect Maxwell's waves. It consisted of a loop of wire with a gap to which small brass knobs had been attached. The loop was connected to an induction coil so that a spark could be generated across a gap. He then built a second loop with an induction coil to act as a detector. When the first loop was connected to the induction coil, a spark jumped across the gap, sending out a "signal." This signal was detected by the second loop (the receiver), which was nearby. Hertz was able to show that the signal exhibited a wave nature, and that it had a certain wavelength, or frequency, so it was obviously another electromagnetic wave. Furthermore, he was able to calculate its speed, showing that it was equal to the speed of light. He announced his discovery in 1887, claiming it was a verification of Maxwell's prediction.

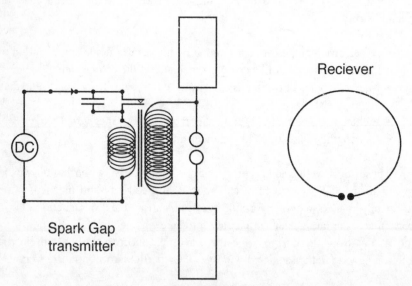

Hertz's apparatus for detecting electromagnetic waves.

THE ELECTROMAGNETIC SPECTRUM

Maxwell was right: there was, indeed, a large spectrum of electromagnetic waves. We now know that they range from very short wavelength (high frequency) gamma rays down to very long wavelength (low frequency) radio waves. In between these two extremes are x-rays, ultraviolet (UV) light, visible light, infrared radiation and microwaves. Furthermore, all of these waves carry energy, or, more exactly, they are a form of energy, and the magnitude of their energy depends on their frequency (number of vibrations per second).[2]

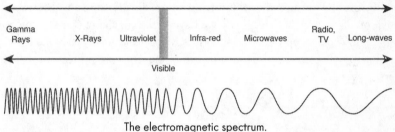

The electromagnetic spectrum.

Several of these types of radiation had already been detected. In 1800 the German-born astronomer William Herschel was studying the temperatures of different colors by moving a thermometer along the spectrum of colors created by a prism when he noticed that the highest temperature was actually beyond red, which was at the end of the spectrum. He concluded that sunlight contained a heat-type radiation that could not be seen. We now know that this is infrared radiation. You can easily detect it when you turn on the burner of an electric stove. You feel the heat long before the burner turns red.

The following year Johann Ritter was looking at the other end of the visible part of the spectrum when he detected invisible rays that were similar to violet rays, but beyond them in the spectrum. He called them "chemical rays," but their name was later changed to ultraviolet rays.

Also, years earlier, in 1895, Wilhelm Röntgen of Germany had noticed a type of high-energy radiation that was created when an evacuated tube was subjected to high voltage. He called the waves x-rays. And Hertz, in some of his earlier experiments, had already discovered microwaves and radio waves. Finally, in 1910 the British physicist William Bragg showed that there were very energetic waves with energies even higher than those of x-rays. These are called gamma rays.

Let's go back now and look more carefully at how we identify each of these types of radiation. As we just saw, they differ in their vibration rate, or frequency. And since frequency is related to wavelength (the distance between equal points along the wave), they also differ in wavelength. In addition, they differ in how energetic they are (we will talk more about this later). For the most part we will identify them by their frequency. The unit of measure for frequency is the Hertz (Hz), which is the number of vibrations per second. The range in frequency, however, is so large that we sometimes have to use units such as megahertz (MHz), which is one million Hertz. Infrared light, for example, has a frequency of up to approximately 100,000 MHz. Microwaves, which you are no doubt familiar with in relation to microwave ovens, have a range from 1,000 to 100,000 MHz. And radio waves go from 1,000 MHz down to 50 MHz. For radiation of even higher frequency we have to use gigahertz (GHz), which is 1 billion Hertz. Infrared radiation is in this range. And finally, beyond infrared, through visible and UV light, a unit called the Terahertz (THz), or 1 trillion Hertz, is used.

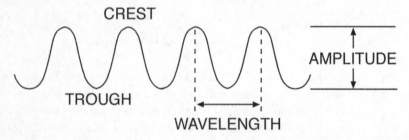

Diagram of a wave showing wavelength and amplitude.

Almost all of these types of radiation have important applications to war. Radio waves are used extensively for communication, and, as we will see, radar played a critical role in World War II, and it is still used extensively. We will also discuss lasers; most lasers use visible light, but other radiation frequencies are now used to produce other types of lasers, and lasers have played an important role in the military. Infrared radiation also has an important application to the military in relation to various devices that are used for night vision. And finally, x-rays are, of course, critical in the treatment of wounded soldiers.

RADIO WAVES

Radio waves are one form of electromagnetic radiation, and it wasn't long after Hertz's discovery that scientists began experimenting with them. One of the first to do so was Guglielmo Marconi (1874–1937) of Italy. When Hertz died in 1894 there was a sudden renewed interest in his discoveries and many newspapers published articles on them. Hertz's work came to the attention of Marconi, who was only twenty years old at the time. He was sure the waves Hertz had discovered could be used to create a system of wireless telegraphy, in other words, a telegraphic system that could send messages without using wires. As a result, he set up a simple system to see if this was feasible. His system consisted of a simple oscillator (a spark-producing radio transmitter) and a "coherer" receiver, which was a modification of an earlier receiving device. He used a telegraphic key to operate the transmitter so that it would send a series of long and short pulses (dots and dashes); a telegraphic receiver was activated by the coherer.[3]

By the summer of 1895 he was able to transmit and receive messages over a distance of a mile and a half. He decided at that point that he would need funding to improve the device. Finding little interest in Italy, he traveled with his mother to England, where he demonstrated his device to William Preece, the chief electrical engineer of the British General Post Office. A series of demonstrations to government officials followed, and with their support, in March 1897, Marconi was able to send a message over a distance of 3.7 miles.

Marconi and his experiments began to attract international attention. In 1899 he set up equipment on two sides of the English Channel and sent a message from France to England. Shortly thereafter he sailed the United States at the invitation of the *New York Herald*. In the following year he began working on equipment to send a message across the Atlantic, and on December 12, 1901, he claimed that he had accomplished his goal. There was, however, some skepticism, so in February 1902, he set up a more advanced apparatus and proved the skeptics wrong. There was no doubt that he had accomplished his goal.

One of the early problems for long-distance radio transmission, or at least anticipated problems, was the curvature of the earth. Since radio waves travel in straight lines, it was expected that this curvature would cut off the signal. Marconi was pleased to find this didn't happen. The reason was that the radio waves were reflected, or bounced, back and forth due to the presence of electrically charged particles in the atmosphere.

Marconi continued to work on his device over the years, but he soon discovered he had competition. His messages used a series of dots and dashes (Morse code), but in the early 1900s, the first vacuum tube was invented, and as a result, wireless voice transmission became possible. This new development quickly overshadowed telegraphic transmission.

Of particular interest, however, was that war departments on both sides of the Atlantic were soon interested in Marconi's device. The British War Office was one of Marconi's first customers, and soon after major German telegraphic firms began buying his products; he set up a company in Germany to start selling them around 1900.

Radio soon started to play an important role in war. Over the next few years transmitters and receivers were improved significantly and radio became the major communication media during war. It was starting to be used by both sides in World War I, and it was, of course, used extensively in World War II.

X-RAYS

Another type of electromagnetic radiation critical in war is x-rays, and its use is mainly for saving lives rather than for ending them. The beginning of x-ray technology goes back even further than that of radio technology. X-rays were discovered by the German physicist Wilhelm Röntgen.[4] At the time, the scientific world was intrigued by a discovery that had been made a few years earlier. High-voltage electrical currents in an enclosed tube containing rarified gases created what were called cathode rays. They had attracted a lot of attention, and Röntgen began experimenting with them. On the evening of November 8, 1895, he discovered something strange. He was particularly interested in a glow, called luminescence, which occurred in certain chemicals, and he wanted to find out if cathode rays would cause luminescence. While working in a darkened room, he noticed that a sheet of paper that he had covered with platinocyanide was glowing. This was strange because the cathode rays were not striking it directly; in fact, they were blocked off from it, yet when he turned the cathode-ray tube off, the glow disappeared. It soon became obvious to him that some type of radiation was being emitted by the cathode-ray tube, but it was invisible, and he discovered that it was coming from the point where the cathode rays were hitting the glass. Checking it out further, he found that this new radiation was highly penetrating. Not only did it pass through wood and thin sheets of metal; it also passed completely through his hand. Furthermore, a photograph using it showed the bones of his hand. He knew immediately that this new type of radiation would have important medical applications, particularly in relation to broken bones, and perhaps in locating bullets and so on within a person's body. He was unsure what to call the waves, so he referred to them as x-rays, and the name stuck. Within a short time (1900) he was awarded the first Nobel Prize for his discovery.

X-rays have, indeed, become an important tool in relation to war. During

World War I, x-ray equipment became a major component of many first-aid stations and hospitals near the field of action. Madam Curie was one of the first to encourage the use of x-rays for treating wounded soldiers in World War I. Over the years, x-ray equipment has improved significantly, and it is now a major tool in war.

LIGHT AND INFRARED

It might seem strange that ordinary light is an electromagnetic wave, but indeed it is. This means that x-rays and light are basically the same; the only difference between them is their frequency. And, as we have seen, frequency is directly related to energy. The frequency of x-rays is much higher than that of light, so x-rays are much more energetic. That's why they easily penetrate your body and can be dangerous.

We can also say that ordinary light is an important weapon of war because telescopes and binoculars have played an important role in war ever since magnifying lenses were invented. The first practical telescope was invented by the Dutch optician Hans Lippershey in 1604, but he kept it secret for several years. Five years later, however, Galileo heard of the discovery and built a telescope for himself.

Simple telescopes have two main lenses: a relatively large convex lens (curved outwardly on both sides) called an objective, and a smaller lens called an eyepiece. Such devices are known as refracting telescopes. A different type of telescope, called a reflecting telescope, uses a mirror rather than a large convex lens. The reflecting telescope was invented by Newton, and it is used mainly in astronomy. Refracting telescopes were used extensively in early wars, and they are still used today. Hans Lippershey also built a binocular version of his telescope in 1608, with two telescopes mounted side-by-side, but it was quite crude. Box-shaped binocular telescopes for terrestrial use were produced in the second half of the seventeenth century and the first half of the eighteenth century by several people, but they were still rather crude.

Modern binoculars use a system of prisms that was discovered in 1854 by Ignazio Porro of Italy. Other systems are also used. Several lenses, in addition to the objective and eyepiece, are now used in modern binoculars.

Turning now to infrared radiation, we find that one of the most useful military applications is the use of infrared for enhanced night vision. Special infrared goggles allow a person to see much better at night. Two types of devices are used. The first uses the infrared wavelengths closest to visible light for image enhancement, using a special tube, called the image-intensifier tube, to collect and amplify the infrared light in this region. (It also gathers some visible light.) A conventional lens captures this light and sends it to the image-intensifier tube, which converts the light signal to electrons with the same distribution. An electron multiplier then increases the beam, keeping the same distribution pattern. These electrons then hit a screen coated with phosphor, which causes them to release a light pattern with the same image as the original one, but enhanced.

The second device uses thermal imaging. It focuses on the infrared region most distant from visible light. In this case a temperature pattern called a thermogram is created. This thermogram is then converted to electrical impulses, and these impulses are sent to a unit called the signal-processing unit, which translates them into a form suitable for display.

Night-vision lenses such as the above are used extensively by the military to locate targets at night. They are also used for surveillance and for navigation.

Lasers, another important recent discovery, also use light in this region, but they will be discussed in a later chapter.

RADAR

Radar is another technology that uses electromagnetic radiation, and, as we'll see in chapter 16, it played a large role in World War II, and it has played a significant role in the military ever since. The word *radar* is actually an acronym for Radio Detection and Ranging. For the most part, it is used for one or more of the following:

- To detect the location of an object that is at some distance away and cannot be observed visually.
- To detect the speed of the object.
- To generate a topological mapping of an area of the earth.

Radar is accomplished using an echo as well as what is called the Doppler Effect.[5] While most people are familiar with echoes, the Doppler Effect is less well known, so I'll explain it. Although radar generally uses microwaves, it is easiest to understand the concepts by explaining them in terms of sound waves. The relevant phenomena are essentially the same. And sound waves are in fact used in the same way as microwaves for what is called sonar. Sonar is important in relation to submarines, and we will discuss it in detail in the next chapter.

Now back to the Doppler Effect. As you no doubt know, sound waves have a certain wavelength, or frequency, in the same way electromagnetic waves do. With this in mind, consider a car approaching you with its horn blaring. The sound wave is traveling away from the car at the speed of sound, but the car is moving, so it is "catching up" with the sound wave. Because of this, the sound wave gets slightly compressed, and this means that its wavelength is shortened (see figure). When the car passes, however, you experience an opposite effect because the two velocities are now in opposite directions. In this case the wavelength is lengthened because the wave is stretched out. As a result, you notice that when the car is approaching, the horn's pitch, or frequency, is higher than it would be if the car were sitting still, and it is lower once the car passes you. The effect was discovered by the Austrian physicist Christian Doppler (1803–1853). It not only occurs with sound waves, but all waves, including radio waves.

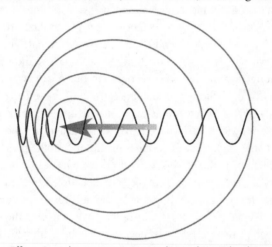

The Doppler Effect. Sound waves are squeezed together in the direction of travel and separated in the opposite direction.

It's easy to understand how we can determine the distance to an object using an echo. If you know the speed of sound in air (it is approximately 1,126 feet per second), and if you measure the time it takes for sound to reach an object and bounce back, the distance of the object can be determined by dividing the time you measured by two and then multiplying the result by the velocity of sound. So the echo gives you the distance. But you can also combine the time of the echo with the Doppler Effect to calculate the speed of an object, such as the car in the previous example. In this case, assume you send a sound signal toward a car that is approaching you. Some of the sound waves will bounce off the car and eventually create an echo, but most will scatter off in other directions. The ones that scatter off in other directions can be ignored. So we will have an echo, but at the same time, since the car is moving toward us, the sound waves will be compressed. The waves of the returning echo will therefore have a higher pitch compared to the original ones. If you measure the difference in pitch of the returning waves, as compared to the ones you sent out, you will be able to determine how fast the car is going. And because the time of the echo gives you the distance, you have both the car's speed and its distance.

In practice, however, sound waves don't work very well. First of all, in most cases the echo would be hard to detect and measure. It would be faint and there would be a lot of interference. Furthermore, sound doesn't travel very far before it decays away. Microwaves don't have this problem, however, and this is why they are used in radar.

So let's set up a simple radar system using microwaves and look at it. Assume that we want to use it to detect enemy planes that are concealed by fog or clouds. First we need to send out a microwave signal. The best form for this signal, as we will see, is a short burst of waves. Assume the burst is one microsecond long (a millionth of a second); in other words, we only turn on the transmitter for one microsecond. This burst will leave the transmitter and travel toward the target; when it reaches it, it will strike the target, and most of it will be reflected. Most of it, in fact, will be reflected in random directions, but some will come back directly to the transmitter, and the pulse that does can be detected and amplified. For this, of course, we will need a receiver, and this receiver is usually (but not always) at the same location as the transmitter. So as soon as the radar transmitter sends out its signal, it is turned off, and the receiver is turned on to listen for the echo. Since radar waves travel at the speed of light, it doesn't take long for the echo to reach the receiver. Immediately after it is picked up, electronic devices measure the time it took for its flight, and they also measure the shift in its wavelength, in other words, its Doppler shift. This information is sent to a computer in the unit that calculates the distance and speed of the approaching airplane, or whatever was being detected.[6]

A radar system can actually detect more than just the speed and distance of an enemy plane. It can also detect the plane's altitude and the direction in which it is flying. And airplanes aren't its only targets. It can be used to detect ships at sea, spacecraft, guided missiles from another country (being shot at us), weather formations such as storms and other phenomena, and the topography of terrain. And it doesn't matter if the targets are obscured by clouds or most other atmospheric phenomena. So radar is obviously a vital part of any defense system.

Let's look at the device in more detail, starting with the transmitter. It sends out microwave radar signals in the direction of the target. Radar signals are reflected by most metals and by carbon fibers; this is why radar is ideal for the detection of airplanes, ships, cars, missiles, and so on. However, radar signals are not reflected well from radar-absorbing materials such as high-resistant materials and certain magnetic materials, and because of this, these materials are used for various types of military crafts and vehicles so that they can avoid radar.

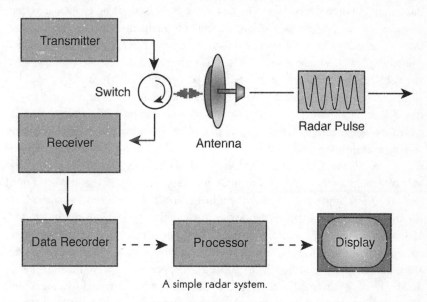

A simple radar system.

And there are also problems in relation to receivers. The reflected microwave beam is usually very weak, so the receiver has to amplify it. Radar beams are, in fact, scattered off a target in the same way that light is scattered off mirrors, but there is an important difference. Ordinary light has a very short wavelength, but microwaves have a relatively long wavelength. And if the radar receiver is to "see" the target properly, the wavelength of the radar signal has to be much shorter than the target size. Early radars used relatively long wavelength signals (in the radio region) and therefore had difficulty in interpreting the reflected signal. More modern radar units, however, now use the relatively short wavelengths of microwaves.

Another problem with radar is that microwaves in the atmosphere and even within the device itself can interfere with the signal. This interference super-imposes itself on the radar signal and has to be reduced or "cleaned up" before the returning signal can be analyzed properly. The interfering waves may come from buildings, mountains, and other objects that reflect microwaves.

AN AMAZING DISCOVERY

In the late 1930s it became obvious that Germany was building up its military and would likely unleash an all-out attack on Britain in the near future. And it was also known that the Germans had close to three thousand planes compared to only eight hundred for Britain. As a result, the British set up an extensive system of radar stations, but radar still had serious problems at that time. It was low-powered and used radio waves that did not give a clear image. The British needed something better, and they needed it fast. The shortest wavelengths available were about 150 centimeters (59 inches) with the power of about 10 watts.[7]

Scientist began a search. It was soon noticed that a General Electric physicist, Albert Hall, at Schenectady, New York, had invented a simple device he called a magnetron in 1920. It looked like it had promise, but he couldn't think of any uses for the device at the time. Hall's device did not generate microwaves, but it was soon discovered that with a slight modification, it might be able to generate them, and, as a result, it attracted some attention. It wasn't until the late 1930s, however, when two engineers in England, Harry Boot and John Randall, decided to explore the device further that people got really excited. Hall's earlier device consisted of a cathode (negative terminal) and an anode (positive terminal) in a glass tube, quite similar to an ordinary vacuum tube. Boot and Randall modified it; they used a copper body, which acted like an anode. It was cylindrical with several cylindrical cavities around its inner edge. These cavities opened into a central vacuum chamber that contained the anode. A permanent magnet was used to create a magnetic field that ran parallel to the axis of the cylinder. The cathode was hooked to a high-voltage power supply. This produced electrons that streamed out toward the cylinder walls. These electrons, however, were deflected by the magnetic field into curved paths, and this caused them to set up small circular currents within the cavities. These currents produced microwave radiation that could be directed into a device called the waveguide, which channeled it to an outside device where it could be used. Of particular interest, the wavelength of the microwave radiation was related to the size of the cavity, and therefore it could be adjusted.

When Boot and Randall completed their device in February 1940, they tested it and were amazed that it produced microwaves with the power of nearly five hundred watts—fifty times what the earlier devices were capable of. Furthermore, the wavelength of the microwave radiation was only 10 centimeters (3.93 inches), which would give a much clearer picture of enemy objects. In addition, the device was small enough to fit into the palm of your hand. They were delighted, and over the next few months they worked to perfect their device.

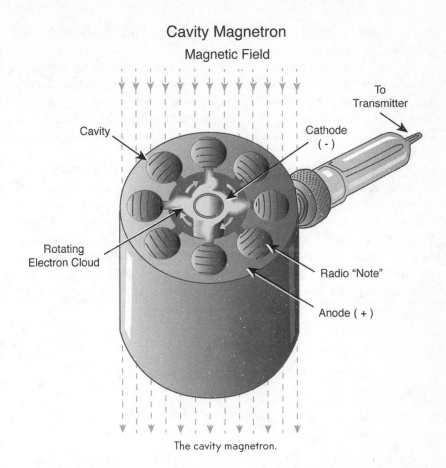

Cavity Magnetron
Magnetic Field

To Transmitter

Cavity

Cathode (-)

Rotating Electron Cloud

Radio "Note"

Anode (+)

The cavity magnetron.

By now, however, the war had started, and Britain was strapped for money. But the British needed the device; in fact, they needed a large number of them for their radar-defense systems against German planes. Churchill knew that Britain could not produce the large numbers needed, but the United States could, and he also knew that the United States was working on its own radar system and would be amazed at the device that Boot and Randall had devised. He therefore suggested that Henry Tizard, the chairman of the Aeronautical Research Committee, offer the magnetron, as it was called, to the United States in exchange for help in mass-producing it.

In a secret mission that took place in September 1940, Tizard went to United States. In a small box he carried a magnetron capable of generating 500 watts (while the most powerful magnetron in United States at the time could create only about 10 watts). And indeed, within a short time, a deal was reached.

American officials later described the device as "the most valuable cargo ever brought to our shores."[8]

Scientists at Bell Labs made a copy of the device suitable for mass production before the end of 1940, and a lab was set up at MIT (Massachusetts Institute of Technology) to develop a more powerful radar system using it. Back in England, scientist at TRE (Telecommunications Research Establishment) developed a revolutionary new radar system that could be used by airplanes for ground mapping.

The magnetron, which is usually called a cavity magnetron because of the small cavities within it, allowed the detection of very small objects such as submarine periscopes. And since magnetrons were now small enough to be installed in airplanes, a squadron of airplanes could easily spot enemy subs and destroy them. The new device also proved valuable in detecting incoming German bombers well before they got to England, so the Royal Air Force could prepare for them. And it also improved the accuracy of Allied bombing raids over Germany. This will be discussed in much more detail in chapter 16.

15

SONAR AND THE SUBMARINE

Ｗe have discussed submarines briefly in previous chapters. In this chapter we will look at the submarine in more detail. It was improved over the years, and it didn't take long for nations to realize that it had considerable military potential.

Although several crude designs for submarines appeared before the 1700s, one of the first to build an operational model was the American engineer Robert Fulton. Between 1793 and 1797 he built the first working submarine while living in France. It could stay underwater for seventeen minutes, and it was about twenty-four feet long. He called it the *Nautilus*. Submarines were also used in the American Civil War. The Confederates built four submarines, the most famous of which was the *H. L. Hunley*. After the war, research on submarines continued, and this research is usually associated with two names: Simon Lake and John Holland. Lake began experimenting with the idea of using buoyancy to submerge and surface a submarine. Holland worked on various methods of propulsion. The US Navy's first commissioned submarine, the USS *Holland*, was built by Holland in 1898. It was fifty-three feet long, weighed seventy-five tons, and it had an internal combustion engine for running on the surface and an electric motor for use while submerged.

All submarines depend on a principle that was formulated many years earlier by Archimedes of Syracuse, Sicily. We discussed it briefly earlier; let's look at it now in more detail.

ARCHIMEDES' PRINCIPLE

Archimedes' principle is related to the pressure on an object in water or another liquid, or more exactly the buoyancy on an object in a fluid.[1] To understand it, let's begin with the concept of pressure; it is defined as force per unit area, or algebraically, as $P = F/A$, where P is pressure, F is force, and A is area. If you're considering the pressure on a given surface under a certain amount of water, it's easy to see that the pressure comes from the weight of the column of water above it acting on the surface. And the weight of this water depends on its density, which is defined as its weight per unit volume. The density of water is sixty-two pounds per cubic foot.

But we're mainly interested in buoyancy, so let's consider a solid cube within a tank of water and determine the buoyant force on it. This is the force pushing it upward. Archimedes' principle states that *the upward force on any object in water or other fluid is equal to the weight of the fluid displaced*. Archimedes (278–212 BCE) arrived at his principle when he was asked by the king of Syracuse to find out if a blacksmith had stolen some of the gold he had been given to make a crown by substituting silver for it. And, as it turned out, he had.

Archimedes' principle is valid if the body is totally submerged, or if it is floating. In fact, it's easy to see that if the weight of the body is less than the weight of the water displaced (when it is totally submerged), the body will float. This means that if its density is less than that of water, it will float.

Let's look now at our solid cube in the tank of water. Assume it is totally submerged. The pressure on all sides will equalize because there is an equal opposing force across from any force acting on it. But the force on the top and the bottom will be different because the upward force on the bottom is greater than that of the force on the top, since the bottom of the cube is deeper. The difference will, in fact, be equal to the weight of the water the cube displaces. This is the buoyant force. And if it is greater than the weight of the cube, the cube will move upward and float. This is, of course, what Archimedes' principle tells us. It occurs for any object in water when the object's density is less than in that of water, and this is, in fact, why ships, which are made of heavy steel, can float. Steel has a high density, but the ship is made up mostly of air, which has a density much less than that of water, so it has an average density less than that of water.

PHYSICS OF SUBMARINES

A submarine can, of course, float on the surface, and it can also dive under the water. And when it is floating its average density has to be less than that of water, but when it dives its average density has to be greater. So it obviously has to change its density, and it does this using ballast tanks that are on its outer surface. When these tanks are full of air the average density of the submarine is less than that of water, so the submarine floats. To submerge, the submarine releases the air through small vents and allows the tanks to fill with water. When they are full (or partially full), the average density of the submarine is sufficient for it to sink. To surface, air is pumped into the ballast tanks from a compressed air tank. It forces the water out.[2]

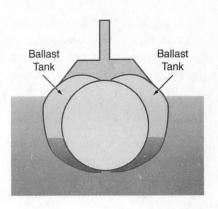

Hydroplanes are also used to assist in the process of diving and resurfacing. They are at the rear of the submarine and look like the wings of an airplane. They help steer the submarine up and down in the same way that rudders on an airplane do.

It is also important to keep the submarine level and steady at various depths when it is underwater. In practice there are several problems. For example, water density increases with depth, so buoyancy increases as depth increases. The temperature of the water also has a small effect. Because of these and other problems, the submarine is in unstable equilibrium when it is submerged, and it therefore has a tendency to rise or sink at one end or the other unless adjustments are made continuously. This is referred to as maintaining trim. To achieve it, submarines use smaller forward and aft tanks. Pumps move water back and forth between them, changing the weight distribution almost continuously. A similar system is also used for stability.[3]

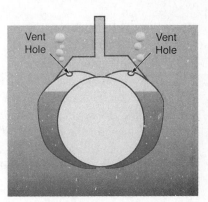

Ballast tanks on a submarine.

POWER FOR THE PROPELLERS

A submarine needs power to turn the propellers, and over the years the power source has changed. The earliest submarines were powered by human muscle. A number of men actually cranked the propeller by hand. But it wasn't long before engines of various types were introduced to do this. By about 1900 gas-powered engines were used on the surface, and electric motors were used when the submarines submerged. Gasoline engines were, however, soon replaced by diesel engines. In the first submarines of this type, the diesel and electric motors were separated by a clutch, so they were both on the same drive shaft to the propeller. This allowed the engine to drive the electric motor as a generator that could be used to recharge the battery that was used for electric power. One of the main problems with the submarines of World War I and World War II was that they had to resurface to recharge their batteries quite frequently. Eventually a snorkel device was invented so they could recharge while still submerged, but they still had to be quite close to the surface.

SHAPE AND PERISCOPES

One of the major problems associated with submarines is hydrodynamic drag. Drag is also a problem in relation to cars, and they are therefore designed with a shape that minimizes it. In the case of submarines, the medium through which they travel is water, and water creates a much greater drag than air does. A teardrop shape for the front was used to keep drag to a minimum on most submarines deployed during World War I and World War II; more recently, however, a slightly different surface shape has been used, although the teardrop shape is still used to some extent.

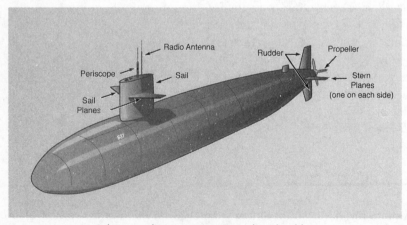

Submarine showing periscope, sail, and rudder.

On the top of the submarine is a raised tower, known as the conning tower, which accommodates the periscope, various electronic devices, and the radio. In many early submarines the control room was also located here. The control room is now located within the submarine and the raised tower is now called the sail. The periscope allows an observer in the submarine to see what is happening on the surface when the submarine is submerged. It consists of a system of mirrors and lenses that bend and reflect images down a long tube. In newer submarines photon masts have superseded periscopes. They are high-resolution, color cameras that send images via fiber optics to a large screen (in fiber optics, pulses of light are sent along a long optical fiber).

NAVIGATION

A modern submarine can use GPS (the global-positioning system) to help guide it while it is on the surface, but when it is submerged GPS does not work. Newer submarines therefore have underwater inertial guidance systems that keep track of their position by noting their motion away from a fixed stationary point. These systems are quite complex, and they generally use gyroscopes to track the location of the submarine. American submarines use a system called SINS (ships inertial navigation system); it keeps track of the location of the submarine by following its course changes using gyroscopes. Numbers are fed to a computer and compared to the starting coordinate. With this system submariners can quickly determine where they are at any time.

Gyroscopes are not only useful for underwater navigation, but, as we will see, they are also used to guide torpedoes to a target. A gyroscope is helpful because it exhibits a fundamental property called gyroscopic inertia, which gives it rigidity in space. As we saw earlier, this is a consequence of Newton's first law of motion, which states that a body tends toward a continuous state of rest or uniform motion unless subjected to an outside force. This means that when a gyroscope is set spinning in a particular direction, it takes a force to change it. Gyroscopes are, in fact, not only used in submarines and torpedoes; they are also used in spacecraft, rockets, guided missiles, and ships, so they obviously play an important role in warfare.

SONAR

Also important in relation to underwater navigation is sonar. When a submarine is submerged it is generally cut off from the region around it because light does not penetrate very far into water. So even if it had video cameras attached to its exterior, they would be of little use. Sonar is similar to radar except that it uses sound waves rather than microwaves. In the last chapter we saw that radar systems send out an electromagnetic pulse then look for an echo, or reflection, of the pulse. By analyzing the echo, radar systems can determine what is around them, even if the objects can't be seen directly. In the same way, sonar allows submariners to see what is in the water around them.[4]

Two types of sonar are used in submarines: active and passive. Active sonar is an analog to radar in that the travel time for a reflected wave is recorded along with any change in frequency of the initial signal. An active signal transmitter, or signal generator, creates a pulse of sound that is referred to as a ping. This pulse is concentrated into a relatively narrow beam, so that it is going in a particular direction. It is used mainly for detecting other submarines, ships, or other objects around the submarine. Analysis of the echo gives information about the distance to the object and the direction and speed at which it is traveling. Its distance can easily be determined from the time between the release of the signal and the return of the echo. Its speed can be determined from the Doppler Effect.[5]

One of the problems with active sonar in warfare is that any ship or submarine in the neighborhood can easily pick it up, which could allow the enemy to determine the vessel's position. Because of this, passive sonar is used in many situations. It is simply a very sensitive underwater microphone that is used to listen for noises in the water around the submarine. The problem, of course, is identifying the sound that the microphone picks up. In most cases, however, this is left to a computer. A large database of different sounds, along with the things that cause them, is stored in the computer. When a particular sound is detected, it is fed to the computer for identification. In general, passive sonar has a greater range, and it has the advantage of being undetectable.

In World War II, active sonar was generally kept to a minimum, so most submarines relied heavily on passive sonar. But modern techniques and devices have improved active sonar, so both active and passive sonar are now used extensively. There are, of course, other problems with both types of sonar. The signal is influenced by the depth of the water and by the water's temperature and solubility, and these factors have to be taken into account. In addition, there is what is called a thermocline in the ocean. It's a sharp division between the warm surface water and the cold, still waters below. As sound waves pass through it they tend to be deflected, and this has to be taken into account as well.

Sonar is used not just by submarines. Objects called sonobuoys were used extensively during World War II, and they are still used. Sonobuoys are small systems, about three feet long and five inches across, which can be easily dropped or ejected from an airplane or ship. They float in the water and can be either active or passive. Their signal is picked up by a nearby ship or an airplane. They do have limitations, however. They have a limited lifetime (depending on their batteries) and a limited range, but they have proven to be useful.

TORPEDOES

Robert Fulton is believed to have been the first to equip a submarine with a torpedo. His submarine, the *Nautilus*, was equipped with a torpedo that was actually little more than a box of dynamite designed to explode beneath enemy ships. He used it in a demonstration in France in 1801 to sink a small ship, and again in a demonstration in England, but he didn't manage to get much interest from either government.[6]

Torpedoes first came into their own during the American Civil War, and they were used most effectively by the Confederate navy. At that time they were mounted on a boom or spar in the front of the submarine, from which they were attached to an enemy vessel. Sometimes they were detonated by the blow they received when they struck the ship, and sometimes timing devices were used. Free-floating torpedoes, or what we today would call mines, were also used. Twenty-two Union ships were sunk by Confederate torpedoes, while only six Confederate ships were destroyed by Union torpedoes. One of the most famous sinkings was achieved by the Confederate submarine *H. L. Hunley*. On the night of February 17, 1864, it rammed and sank the Union ship USS *Housatonic*, but the explosion was so great that it damaged the *Hunley*, and it also sank with all men aboard. In 2004 the remains of the *Hunley* were located and raised.

One of the most significant advances in torpedo technology came in 1864. An Englishman, Robert Whitehead, who was working in Austria, became interested in the torpedo and decided to build a model that would run just under the surface of the water. In October 1866, he had his model ready. It was driven by a two-cylinder compressed-air engine, had a top speed of about seven and a half miles per hour, and a range of approximately two hundred yards. Officials in Austria were so impressed that they immediately purchased it. He did, however, also sell the rights to manufacture it to several other countries. Strangely, the US Navy was not interested in it.

HOW TORPEDOES WORK

A modern torpedo is a self-propelled projectile. It is stored in the launching area until it is fired. When fired, it is given an initial velocity, but as it moves out into the water several other forces act on it. Gravity pulls it downward, and the drag of water on it creates a friction that slows it down. The friction created by the drag of water is, indeed, quite large—about a thousand times greater than the drag created by air. Depending on design, there is another force that also comes into play, namely the buoyancy of the torpedo itself. All of these forces have to be taken into consideration.

The earliest torpedoes used compressed air to turn their propellers. Within a few years, however, it was found that compressed oxygen was more efficient, but oxygen systems posed a danger to submarines when they came under attack. Because of this, the Germans used a small electric motor powered by batteries. It had an additional advantage in that bubbles were not released as a torpedo moved toward its target. It was slower and had a more limited range than previous torpedoes, but it was much cheaper to build. The United States also soon introduced an electric motor model called the Mark 18.

Torpedoes can be aimed at a target and fired in the same way that an artillery shell is fired. In this case, there is no control over the torpedo once it leaves the submarine, and no changes can be made if the target sees it and tries to outmaneuver it. Because of this, guided torpedoes are frequently used. In some cases they are guided to the target by its sound or by the use of sonar. They are referred to as acoustic torpedoes. The first acoustic torpedoes were employed late in World War II by the Germans, and they proved to be quite effective against both surface ships and other submarines.

Acoustic torpedoes are equipped with acoustic sensors and transmitters in their nose. They are therefore able both to detect sound coming from the target and to produce sonar reflections. Usually the torpedo starts by using passive sonar. Once the passive sonar has detected the enemy, it switches over to active sonar, which allows it to send out a sound beam to locate the enemy more exactly. It then attacks.

Another effect that is useful in the case of torpedoes is what is called supercavitation. When an object moves through water at high speed, the pressure behind the object is lowered, and, as a result, a bubble is formed that can encompass the object. This is particularly useful in the case of a torpedo because water creates a large frictional drag on it. If the object is in such a bubble, however, the drag is significantly reduced. Torpedoes are therefore designed to produce supercavitation bubbles.

SUBMARINES IN WORLD WAR II

Submarines were used extensively in World War II by both sides. They were particularly effective for the Germans at the beginning of the war and for the Americans against the Japanese toward the end of the war. Although they were limited in speed, range, and endurance while they were underwater, they could attack with total surprise and inflict devastating damage, so they were highly lethal. Although early submarines were basically underwater crafts, they spent a lot of time on the surface, submerging only when they were engaged with the enemy.[7]

The Treaty of Versailles that ended World War I did not allow Germany to build either surface ships or submarines. But the Germans soon found that they could build submarines much more quickly and with more secrecy, so they concentrated on them, and by the beginning of World War II they had the largest fleet of submarines in the world. Furthermore, they had also now developed several new technologies and techniques, so their submarines were better than the British and American ones. The early U-boat success of the Germans was mainly due to a World War I U-boat captain Karl Doenitz. He built up the submarine fleet and equipped it with highly trained crews, and he developed what was called the Wolfpack tactic, which was particularly effective. To initiate the tactic, German U-boats would spread out across a large section of the ocean looking for convoys. When one was located, the captain of the U-boat that located it would signal the other U-boats, and they would group themselves around and ahead of the convoy. All convoys were escorted and protected by destroyers and other battleships at that time, so the U-boat captains had to outsmart the escorts. They would therefore attack together at night, creating as much chaos as possible; this would give the German submarines a much better chance of escaping. Allied losses were high when this tactic was first used.

The main objective of German U-boats was to cut off supplies to Britain from the United States. And the U-boats sank so many merchant ships that it appeared they might achieve their objective. There was a problem, however: members of the German high command, and Hitler in particular, didn't realize how effective their submarines were. Hitler was concentrating on the land war, and submarines were not high on his list of priorities. So when Doenitz asked for more U-boats, Hitler turned him down.

Eventually the British and Americans developed very effective antisubmarine devices and techniques. And by 1943 the Wolfpack tactic was losing its effectiveness. Radar and sonar were being used along with a particularly useful surface technique called Huff Duff, in which radio sources were pinned down by triangulation. In addition, British code experts finally broke the code the German U-boat captains were using, and, as a result, they knew when and where to expect an attack. In May 1943, Doenitz lost forty-one U-boats in three weeks, and the tide began to turn. Until the end of 1942 the Germans had been sinking fourteen ships for every submarine loss, and now they were losing their submarine fleet at a crippling rate.

U-boats still had to come to the surface to recharge their batteries, and they were particularly vulnerable above water now. They could easily be located by radar and Huff Duff, and airplanes would be quickly dispatched to attack them. As a result, they were forced to stay submerged most of the time. The Germans eventually developed a snorkel so that they could recharge their batteries while just under the surface, and this helped. But in the end, Germany lost almost 80 percent of its U-boats. In addition, British submarines began sinking German submarines at a high rate; in all, they sank thirty-nine.

Submarines were also used effectively in the Pacific Ocean by the Americans against the Japanese. America entered the war following the surprise attack on Pearl Harbor by the Japanese. In two hours, Japanese pilots killed two thousand four hundred people and wounded another seven hundred. And although they crippled the US fleet, they ignored the nearby submarine base and the fuel and ammunition supply depots. In addition, all the large US aircraft carriers were at sea during the attack. For this reason, the main American fighting force that was left consisted of aircraft carriers and submarines. Unfortunately, neither the submarines nor the torpedoes they carried were a match for the Japanese and German submarines.

At the beginning of the war the American submarines had no radar, and they were armed with relatively poor torpedoes that frequently misfired. Nevertheless, they fought back quickly with what they had. One month after the attack on Pearl Harbor the USS *Pollock* sank a Japanese freighter near Tokyo Bay. And relatively early in the war the United States made an important breakthrough: cryptanalysis groups deciphered Japanese coded messages. So they knew what the Japanese were planning—their strategies and movements. And this, of course, proved to be extremely helpful.

Although the American submarines were inferior at the beginning of the war, a tremendous effort went into improving them, and by August 1942, the first radar systems were installed in US submarines. Then new Gato-class submarines began replacing the older submarines. After many embarrassing experiences with their torpedoes, they were also gradually perfected, and new tactics were developed. In 1943 they even started copying the German Wolfpack tactic, but it didn't work as well against the Japanese.

Although the Japanese submarines and torpedoes were superior at the beginning of the war, the Japanese never took advantage of it. They used their submarines mostly against American warships, which were much more difficult targets than merchant ships. And they were generally used in connection with groups of warships.

By the end of 1943, American submarines were beginning to inflict a tremendous amount of damage to Japanese naval forces. Although the submarine fleet constituted only 2 percent of the overall American navy, it destroyed 30 percent of the Japanese navy, including eight aircraft carriers, a battleship, and eleven cruisers. It also destroyed 60 percent of the Japanese merchant fleet, which created a real problem for Japan. It quickly began to run out of resources.

Submarines were particularly effective in the Battle of the Philippine Sea in June 1944, when they sank two of the largest carriers in the Japanese fleet. Shortly after 8:00 a.m. on July 19 the USS *Albacore* sighted the *Taiho*, the largest and newest carrier in the Japanese fleet. As the captain of the *Albacore* started to fire torpedoes, however, its torpedo data computer failed, and the torpedoes had to be fired manually by aiming them as accurately as possible. Six torpedoes were fired; four of them missed and a Japanese pilot spotted one and crash-landed his plane into it before it reached the carrier. But the last one got through, striking the carrier on the starboard side, rupturing two of its aviation fuel tanks. At first the damage did not appear to be serious, but it produced a lot of explosive fumes. And an inexperienced damage-control officer ordered that the ventilation system be turned on full blast to clear out the fumes, but this only spread them, increasing the danger of ignition. Hours after the initial torpedo impact, several large blasts occurred. Before long the carrier was underwater.

A few hours later the submarine USS *Cavalla* sighted the carrier *Shokaku*. It fired six torpedoes at it, hitting it with three. One of the three hit its forward aviation tanks, causing them to explode. Soon there were other explosions, and within minutes the entire carrier was in flames, and it quickly rolled over and slipped beneath the waves.[8]

This battle was one of the last for the Japanese navy; it was never able to recover from the serious losses it incurred.

After World War II, a new and vastly improved submarine came into use, the nuclear submarine, which we will discuss in a later chapter.

16

THE GREAT WAR

World War II

World War II was the deadliest and most destructive war in the history of the world. A large number of nations—fifty in all—took part in it. And no previous war had such a profound effect on physics and science in general. A large number of weapons were developed and improved over the six years of the war, and a number of important new innovations came about as a result of physics:

- Advanced techniques in radar.
- Rockets, including the V-1 and the V-2.
- The first jet airplanes.
- Code-breaking computers and other computer developments, including ULTRA, ENIAC, and COLOSSUS.
- Proximity fuses and many other guided-artillery devices.

Physics and other sciences were mobilized on a large scale during World War II, with large research and development labs directed mainly at war projects coming into being for the first time. Some of the larger ones were the radiation lab at MIT, Bletchley Park in England, the Los Alamos lab, and the Manhattan Project, and of course the Germans also had their research labs.

Physics became critical in relation to accuracy of artillery weapons, strategic bombing, navigation of airplanes, ships, and submarines, and the development of radar, to mention only a few. It was the first "high-tech" war, and it was largely fought using new technologies, many of which came from physics.

HOW THE WAR STARTED

Historians generally agree that World War II was primarily a result of the surrender conditions that were imposed on Germany at the end of World War I. But it was also due to the economic conditions that soon came about after the war. Unemployment was high throughout Germany, and inflation made German money almost worthless. Exacerbating this was the worldwide depression that hit in 1930. Trade fell off and millions of workers throughout the world lost their jobs. The economic situation was particularly bad in Europe, and people began to look to their leaders for change. They wanted relief.[1]

During this time several dictators came to power. Mussolini and his Fascist Party took over Italy in 1922, and in the early 1930s the Nazi Party under Hitler began to rise in Germany. In addition, military leaders began to control Japan. As a result, all these countries came under the control of totalitarian governments that did not allow any opposition. Furthermore, these leaders promised great things; they assured their people that the nation would be great again and that there would be a significant turnaround. And the people believed them.

Soon after he gained almost complete control, Hitler was already thinking of retaliation against France, England, and other nations. He was bitter, and he had many bitter followers. He was hindered by the Treaty of Versailles, which prohibited Germany from organizing a large army or rearming in any significant way. But Hitler wasn't interested in complying with it, and he soon entered into a coalition with Russia. This arrangement allowed him to manufacture military weapons deep in Russia, away from the prying eyes of inspectors. In return, he gave Russia many military secrets for new technical devices. Not only were German tanks and airplanes being built deep in Russia, but training camps were also set up to train pilots and develop a new, highly skilled army. Much of the manufacturing was done by the giant German steel company Krupp. Finally, by about 1935, Hitler gave up all secrecy and began manufacturing directly in Germany. He almost dared England and France to try to stop him.[2]

During this time England and France were also mired in the depression, and they were directing very few resources to the military. It was a cooling-off period for them, but they were becoming worried about Germany.

READY FOR WAR

By the late 1930s Germany had rebuilt itself into a major military power, unmatched by any other country in Europe. Airpower was a major concern to

Hitler. World War I had been fought mainly in the trenches, and for several years it had been a stalemate, with neither side making significant advances. Hitler was determined to overcome this; he didn't want a repeat of World War I. New techniques and strategies were therefore needed, and one of the main parts of his new strategy was a large and highly efficient air force (the Luftwaffe). And by the late 1930s Germany had an air force unmatched in Europe. Not only were German planes superior to those of both England and France, but Germany also had a much larger number of planes: 5,638 fighters and bombers against 1,070 for the English and 1,562 for the French. In addition, the pilots in England and France had grown rusty, but German pilots were highly trained and had seen action in the Spanish Civil War between 1936 and 1939. Furthermore, Germany had a large number of long-range bombers that were a significant threat to England.

The German tanks were no better than the French tanks, but the Germans had many more of them, and all of them were equipped with radios. And they had beefed them up so that there was little defense against them. Antitank weapons existed, but there were so few of them available that they would be of little help. Most of the opposing bullets bounced off the tanks and did little to stop them. Germany's biggest advantage at the time was a new strategy called *blitzkrieg* (German for "lightning war"). It was based on fast movement using a concentrated tank and airplane attack along with fast-moving troops. The idea was to continue the attack regardless of the opposition; in short, stop at nothing. And it worked well. German tanks were relatively fast and almost impervious to artillery, and they were accompanied by an all-out air attack, mostly by dive-bombing Stukas. The Stuka was quite effective during the early part of the war; it delivered its bombs as it dove down toward its target, and because of this it was highly accurate.

Hitler began by attacking Austria. He had always seen Austria as part of Germany, as he had been born in Austria, and he wanted to annex it. But the Treaty of Versailles forbade a union between the two countries. Nevertheless, he invaded Austria in March 1938, and to his delight there was little opposition. Large crowds actually cheered him as he entered in triumph. Austria became part of the German Reich. Next came Czechoslovakia, beginning with the annexation of the northern and western border regions. These regions contained many Germans, and Hitler's pretext for occupation was that he was freeing them from suffering under the Czech government. In March 1939, Hitler moved his troops into the remaining parts of Czechoslovakia, which was now weak and almost powerless to resist. Although the Czechs fought bravely, they were no match for the Germans and were quickly overcome.

Hitler then looked toward Poland, but there were several problems. Poland had an agreement with France and England, and he had no excuse for an invasion. Poland had been arming itself for such an invasion, but it was no match for the German military in numbers or in tanks or airplanes. Furthermore, just prior to the invasion, Hitler had signed a pact with the Soviets. They agreed to remain neutral if France and England entered into the war. In return, Hitler agreed to share Poland with the Soviets after it was conquered.

The attack began with several events staged by the Germans to use as an excuse for the invasion. During the night of August 31, the Germans staged a fake attack on a radio station near the border, using Germans posing as Polish troops. The next morning Hitler ordered an attack on Poland without a declaration of war. The German Luftwaffe attacked the Polish town of Wieluń, killing close to twelve hundred people, mostly civilians, and leveling about 75 percent of the city. Within a short time German troops attacked the western, southern, and northern borders, as the Luftwaffe began bombing major Polish cities. Their blitzkrieg tactic was used effectively in the attack. Polish forces were quickly forced back from their positions near the border as the Luftwaffe bombed Polish airstrips and many early warning sites.

Within two days of the initial attack, France and England declared war on Germany. The Poles hoped they would soon get aid, but very little came. Polish forces managed to hold off the Germans for two weeks, but then the Soviets invaded Eastern Poland on September 17. The Poles now had to fight on two fronts, and they could do little to stop the two advancing armies. By October 6, German and Soviet forces controlled most of the country. Surprisingly, though, the Poles never formally surrendered. They soon organized an extensive underground force that continued to fight the Germans for years.

THE BATTLE OF FRANCE, AND DUNKIRK

After the Polish invasion, Britain deployed troops to the continent, mainly France, but little happened. Neither side attacked, and for several months both sides waited. It was later referred to as the "phony war" by the British and the "sitting war" by the Germans. Then, in April 1940, Germany invaded Denmark and Norway. Denmark succumbed almost immediately and Norway was overcome in about two weeks. Still, England and France did nothing. Then Winston Churchill became prime minister of England.[3]

On May 10, 1940, the stalemate was broken when the Germans invaded France, Belgium, the Netherlands, and Luxembourg. Employing the blitzkrieg tactic that they had used earlier, they overran the Netherlands in a few days and Belgium in a few weeks. The French, however, were a more significant foe, and they also had help from a relatively large British force, so it was expected that they would stop the German attack, but they didn't. The German army quickly burst through the Ardennes and moved rapidly to the west before turning northward toward the English Channel, reaching it on May 20. The German spearhead separated the British and French forces, backing them up against the sea. It appeared that the Germans might trap and capture them. But surprisingly, the Germans stopped their advance at this point for about three days to regroup and plan their next move. This gave the Allies time to prepare for an evacuation across the English Channel. There were far too many soldiers for the ships available, but soon after news of the situation reached the English public, a large flotilla of merchant ships, fishing boats, pleasure boats, and other craft raced across the channel to the beach at Dunkirk. And over the next nine days, 380,226 English and French soldiers were rescued and brought back to England. They were attacked from the air by German planes as they boarded, but the loading continued. Finally, they had to evacuate under the cover of darkness, but the vast majority of the stranded soldiers made it out safely. Two French divisions remained behind to protect the evacuation, and although they slowed the German advance, they were eventually overcome and captured. The remaining French army surrendered on June 3, and the Germans marched into Paris on June 14. The formal surrender was on June 22.

THE RADAR ADVANTAGE

Both sides had developed radar by the outbreak of World War II, but the British took advantage of it to a much greater degree. They developed it extensively during the early part of the war, and they used it much more effectively than the Germans. The Germans, in fact, underestimated its possibilities and never took it seriously. But they found it to be a serious problem when they attacked England, and indeed it played a large role in the British victory in the Battle of Britain.[4]

When researchers started working on the new technique, the term *radar* was not yet in use. The technology was referred to as RDF, meaning range and direction finding. The earliest research was initiated in 1935 by the British Aeronautical Research Committee, which was headed by Henry Tizard. By this time Germany was making no secret of its military buildup, and many in Britain were beginning to get worried. A research project for detecting incoming planes—particularly German bombers—was initiated by Robert Watson-Watt. His team soon showed that a plane could be detected at a distance of seventeen miles using reflected radio beams. And soon an extensive RDF program was in full swing.[5]

In 1936 the program was moved to Bawdsey Research Station in Suffolk, with Watson-Watt as its director. With a team of many of the best scientists and engineers in England, Watson-Watt improved the technology significantly. And within a short time, a chain of radio stations was constructed along the south and east coasts of England. It was referred to as the Chain Home, or CH system. It was relatively simple, and because it used radio waves from ten to fifteen meters long (20 to 30 MHz), the images that were received were rather fuzzy. And at the time, oscilloscopes were used to display the images. It was crude, but with a little work the operator could determine the direction and approximate altitude of an incoming bomber.

Using this system, British operators could "see" incoming German bombers and send fighter planes out to encounter them. It was particularly helpful in that these planes could be dispatched only when they were needed so that they didn't have to waste a lot of fuel patrolling the English Channel.[6]

It didn't take long for the Germans to realize that the British were detecting their airplanes, so early on they tried to bomb some of the visible radio towers, but they were not very successful. Even when they disabled a particular system, it was usually back in service within a few days. As a result, they soon shifted to a new tactic. They decided to fly at very low levels, under the line of sight of the CH stations, but the British had another system called the Chain Home Low (CHL) system, which had originally been developed for another purpose (naval guns), and it was able to detect the incoming German aircraft.

A new and significantly better system was developed and put into operation in January 1941. It was called the ground-controlled intercept (GCI) system. In this system the antenna was rotated, allowing for a two-dimensional representation of the airspace around the operator. Incoming aircraft appeared as bright dots on a screen, similar to what we see today. The indicators, which were called plane positive indicators (PPI), were a significant improvement over those used in CH. The position and altitude of a plane could now be determined quickly.

Then, in 1939, Edward Bowen and his team developed a small radar system that could be used in airplanes and submarines. It was called the Air Interception (AI) system. It was quickly placed in many of the British aircraft and submarines, and it gave an even better fix on incoming German planes. The Germans tried to avoid it by flying only at night and in poor weather. This had no effect on the radar system, although the British pilots sent out to attack them had more trouble locating them.

In early 1940, however, the cavity magnetron was invented by John Randall and Henry Boot, and it revolutionized radar. It had been known for years that if short-wavelength radiation could be used in radar systems, it would significantly improve the images. But as they decreased the wavelength, the power of the system decreased. This changed with the introduction of the magnetron. With it, "centimeter" radar was possible for the first time, and the power of the new units was much greater. There was still a problem, however; large numbers would be needed, and Britain was in no position to develop and produce them. As we saw earlier, this led to the Tizard Mission in September 1940, which resulted in the manufacture of a large number of magnetrons in the United States.[7]

High-quality radar units could now be mounted in airplanes, ships, and submarines, and they were extremely effective. With their high resolution they could detect objects as small as submarine periscopes. And this quickly blunted the effectiveness of the German U-boat program. Large numbers of U-boats were hunted down and sunk, until finally the Germans withdrew their fleet.

Further advances in radar continued throughout the rest of the war. In 1938 the coastal defense (CD) system was developed. It was not an airborne radar unit, so it could be made much more powerful.

Several defenses against radar were developed during the war, and they were used by both sides. Radar "jammers" were developed that transmitted radio signals of the same frequency as the radar. The jammers were used to saturate the receiver with strong signals so that it couldn't detect properly. "Chaff" was also used. It was a cloud of lightweight strips of metal of a specific size that potential radar targets could deploy to interfere with the incoming signal. The radio receiver would only see a huge cloud when it was used.

But luckily for the British, the Germans didn't take radar very seriously, and they never put a large effort into developing it or protecting against it.

THE BATTLE OF BRITAIN

Soon after France surrendered, Germany turned its attention to England, and what took place over the next couple of months was one of the most famous battles of World War II. And it was totally an air battle. Hitler knew that for total control over Europe he would eventually have to attack England. In particular, he would have to land large numbers of troops on British shores. But while they were landing they would be under constant attack from Britain's powerful navy and also from its air force. He knew that the losses would be staggering and that he had to disable the RAF (Royal Air Force) and the navy before beginning his Operation Sea lion, which was his code name for the land invasion of Britain.[8]

Hermann Göring, the commander-in-chief of the Luftwaffe, assured Hitler that his airplanes could defeat the RAF in the south of England within four days and destroy the rest of the RAF within four weeks. Hitler was therefore overcome with confidence, and he scheduled Operation Sea lion for September 15. And there was no doubt that Germany had an overwhelming superiority in numbers: over 4,000 aircraft compared to Britain's 1,660. Included in this number for the Germans were 1,400 bombers, 800 fighter planes, 300 dive bombers, and 240 twin-engine fighter bombers. The RAF had mostly Spitfires and Hurricane fighters.

The battle began on July 10, 1940, with the Luftwaffe bombing coastal shipping centers and ship convoys. But by the end of July, British fighters had shot down 268 German planes and had lost only 150. As a result, the Luftwaffe switched to attacking airfields, operation rooms, and radar stations. They hoped, in particular, to disable the British radar system. The Stuka had been used extensively in the German blitzkrieg attacks, and it had been particularly effective when used over Poland, France, and Belgium, where there was little defense against it. But it had never encountered a fighter like the Spitfire, which had a top speed of 350 miles per hour compared with the Stuka's top speed of approximately 200 miles per hour. Furthermore, the Stuka was much less maneuverable, and its ability to dive was not an advantage in an air war. By the middle of August, nearly all the German Stukas had been destroyed by Spitfires and Hurricanes. Göring quickly pulled the few remaining ones out of the war.[9]

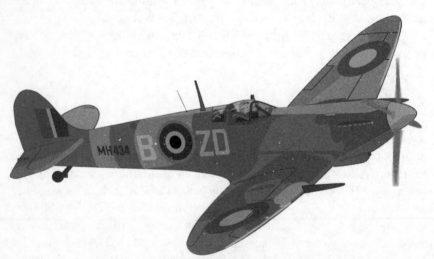

Britain's big advantage (aside from radar): the Spitfire.
It was faster and more maneuverable than most German planes.

Germany may have had an advantage in numbers, but Britain had several important advantages. German fighters barely had enough fuel to get to England and back; they were used to protect the bombers, but they had to return soon after the bombers were over England, and this left the bombers vulnerable to attack. As a result, large numbers were shot down. Furthermore, German fighters frequently ran out of ammunition over England and had to head for home quickly. British planes could easily land and quickly reload their guns. And of course the biggest advantage the British had was their radar detection system. The RAF therefore knew at all times where the German bombers and fighters were, but the Luftwaffe pilots could only guess where the British were.

After August 23 the Luftwaffe stopped attacking seaports and radar sites and switched to night raids on cities—London, in particular. But the Germans continued losing planes at a rate of almost two to one: 1,000 German planes to 550 of the RAF's planes. And on September 15 the Luftwaffe lost sixty planes to the RAF's twenty-eight in a single day. Two days later Hitler postponed the invasion of Britain indefinitely. But the indiscriminate bombing of the larger cities continued. In the end, both sides had taken heavy losses, but the German losses were much greater. And finally, about the middle of October, the raids ceased except for an occasional bombing. The Battle of Britain was over and the British had won. But the war was far from over.

AMERICAN ENTRY INTO THE WAR

The United States entered World War II following the Japanese bombing of Pearl Harbor, but even before 1939 most Americans realized they would eventually be involved in the war in Europe. The trigger, however, came on December 7, 1941. Over a period of four hours six Japanese aircraft carriers sent waves of torpedo planes, fighters, and dive bombers over Pearl Harbor. Given the simmering political tensions between the United States and Japan, there was some expectation among American military leaders that the Japanese might attack. Nevertheless, the US forces were caught completely off guard. As a result, the Japanese planes were able to destroy or severely damage eight battleships, ten smaller warships, and two hundred thirty aircraft, while killing 2,400 US personnel.[10]

The following day the United States declared war on Japan, and since Hitler and Mussolini had just signed a pact with Japan, Germany and Italy declared war on the United States. Although Japan didn't follow up its attack on Pearl Harbor, it did attack the US air base near Manila in the Philippines, and the

Japanese army invaded and trapped large numbers of Americans and Filipino forces in nearby Bataan, leading to the infamous Bataan Death March, during which thousands of Americans and Filipinos died. General Douglas MacArthur escaped to Australia and vowed to return. The Japanese continued their invasion, seizing the Dutch East Indies, and then the islands of Tulagi, Guadalcanal, and the Solomon Islands. They seem almost unstoppable.

What was left of the United States Navy first met the Japanese in the Coral Sea near the Solomon Islands. During two days of combat the Japanese lost a small carrier, a destroyer, and several smaller ships, but the United States lost a carrier and a destroyer, so the battle has generally been considered a draw. But in the process the Americans stopped the invasion of an island that would have allowed the Japanese to strike Australia. It was also an important battle in that the Americans learned a lot about Japanese tactics, and this would help them later in the war.

One of the major naval battles of the war came in June 1942. The Japanese admiral Yamamoto was planning a large offensive near Midway Island. He hoped to trap and destroy most of the American fleet in a quick and decisive battle, but American intelligence, which had managed to decode Japanese messages, knew what he was planning. This allowed American admiral Chester W. Nimitz to set up a web of decoy tactics and plan an ambush. And when the battle ended, the Japanese had lost four carriers and all the airplanes that had attacked Pearl Harbor, along with a large number of Japanese pilots. In turn, the Americans had lost only one carrier. This was a major defeat for the Japanese, and a turning point in the war in the Pacific. The American navy now had clear-cut superiority over Japan's navy.

The battle of Leyte Gulf in the Philippines, which came in October 1944, was one of the largest naval battles in history. It was also a decisive victory for the US Navy, which sank most of the remaining Japanese fleet. What remained of the Japanese navy finally retreated back to Japan.

Over the previous few years Japan had occupied a large number of islands in the South Pacific, and American forces began a strategy known as island hopping, in which they targeted islands that could support airstrips that would allow them to get closer and closer to Japan itself. At the same time they applied their air power to cut off all supplies to Japanese troops on the various islands. The Japanese, however, had dug in, and many of them were in bunkers and caves. Furthermore, American marines soon found that most Japanese preferred to fight to the death rather than be captured. So the fighting was difficult.

Hand-fought battles at Guadalcanal, Tulagi, the Marshall Islands, Iwo Jima, and Okinawa followed. In most cases the Japanese fought until the last man was killed. Furthermore, Japanese pilots were now flying kamikaze missions in which they would commit suicide by flying their planes into American ships. In this way they managed to sink thirty-eight ships and damage many others.

Because of this, the American high command decided that an invasion of mainland Japan would lead to the loss of too many American lives, with the Japanese refusing to the end to give up. President Truman therefore ordered the dropping of atomic bombs on Hiroshima and Nagasaki in August 1945, and within a short time the Japanese surrendered. We will discuss this in more detail in the next chapter.

Turning now to the war in Europe, the first American operation was in November 1942, when American and British troops landed in North Africa. They stopped the German advance on Tunisia, and by May 1943 they had defeated the Germans, capturing over 275,000 men in the process. Along with the British, they then turned to what they believed to be the weakest link of the German and Italian defense, namely Sicily. In July 1943, a large amphibious invasion was unleashed, and in a little over a month Sicily was under Allied control. The Allies then turned their attention to the Italian mainland. American troops landed in Italy in September, and Italian troops surrendered almost immediately, but a large number of German troops were now in Italy, and they continued to fight through the winter. But in June 1944, Rome fell, and soon the Allies had occupied most of Italy.

Meanwhile, in England the largest amphibious attack in history was being planned. The operation, which began on June 6, 1944, consisted of 4,600 ships and over a million troops. Under the command of General Dwight D. Eisenhower, the Allies crossed the English Channel in an effort to establish a beachhead in Nazi-occupied France. The Germans had been expecting an invasion, but they didn't know where the Allied forces would land. For two months preceding the invasion British-based aircraft had bombed airfields, bridges, and rail lines throughout France. And on the night before the landing, paratroopers were dropped inland as naval guns powdered installations along the shorelines. The various landings were given code names; the British and Canadians landed at Gold, Juno, and Sword Beaches, while the Americans landed at Utah and Omaha Beaches. The Canadian and British landings went relatively smoothly, meeting little opposition, but the Americans were met by heavy German gunfire that inflicted many casualties. Within five Days, however, sixteen Allied divisions were in Normandy, and the drive to liberate Europe was under way. By August 25 Paris was captured, as the Allies continued their push toward Berlin.

In the east, the Soviets, who had beaten back a German invasion, were also pushing toward Berlin. Although it was now almost certain that Germany would soon be defeated, the Germans didn't give up easily, and in December 1944 they launched a massive counterattack in the Ardennes Forest that caught the Allies off guard. This engagement became known as the Battle of the Bulge because of the large bulge it created in the Allied lines. By late January, however, with large numbers of Allied reinforcements rushing to the front, the German offensive was stopped. And in March, Allies crossed the Rhine River and began a final push toward Berlin. The remaining German forces were now being squeezed from the east and the west. On May 2, 1945, the Germans surrendered.

ADVANCES IN AVIATION

Let's go back now and look at some of the important advances that were made during the war, many of which depended on physics. Major advances in aircraft design occurred, with the most important being the building of the first jet plane. Aside from the first jet plane, however, there were significant advances in traditional aircraft. Let's begin by looking at some of the major planes that were used in the war, and there capabilities. The British Spitfire was, without a doubt, one of the best. It was used very successfully against the Luftwaffe in the Battle of Britain. It had a maximum speed of approximately 350 miles per hour, and it performed well in climbs; furthermore, it was relatively easy to fly. The British Hurricane was also an excellent plane, and it was also used extensively in the Battle of Britain.[11]

The German Messerschmitt 109 was the only German plane comparable to the Spitfire. It had a maximum speed slightly less than that of the Spitfire, and it was less maneuverable, but it was faster in a dive.

The Japanese Mitsubishi Zero was the primary Japanese naval plane. It was used in the attack on Pearl Harbor and throughout the Pacific war. In the early years no American plane was a match for it. By late in the war, however, it was no match for most American planes.

The P-51 Mustang was one of the best American planes. It had a top speed of 370 miles per hour and was a favorite among American pilots. Many considered it the best fighter plane of the war. Its speed, maneuverability, and range made it an excellent aircraft. Another of the American planes was the Lockheed P-38 Lightning. It is said to have shot down more Japanese planes than any other American fighter during the war. It had a top speed of 414 miles per hour. Another excellent American plane was the F4U Corsair, which was used by US naval and marine pilots. It was the first plane to finally give Americans superiority over the Japanese zero, as it was much faster and had a better roll rate. Its maximum speed was 435 miles per hour.

The fastest and most interesting plane of the war, however, was the Messerschmitt Me 262, which was the world's first jet plane. It had a maximum speed of about 530 miles per hour, which was 93 miles per hour faster than the swiftest Allied fighters. Fortunately for the Allies, it came into the war relatively late, and only a few were built, so it had little impact. Nevertheless, German pilots of the Messerschmitt Me 262 shot down approximately 540 Allied planes, and they were so fast that they were difficult targets. They were so fast, in fact, that German pilots had to learn new tactics when using them in combat. Allied pilots soon found that the best way to deal with them was to attack them on the ground or during takeoff or landing. Airfields in Germany that were identified as jet bases were therefore heavily bombed. The Me 262 did have a number of drawbacks, however; it used twice as much fuel as a conventional aircraft, and near the end of the war, Germany was running short on fuel. Furthermore, there were engine reliability problems.

The jet engine was invented by two different inventors at about the same time: Hans von Ohain and Frank Whittle. Frank whittle was the first to patent a turbojet engine; in fact, his patent came in 1930, six years before Ohain's. But neither man knew anything about the other's work. But it was Ohain who was first to build a workable jet plane.

Whittle was a pilot and an English aviation engineer who joined the RAF in 1928. At the age of twenty-two he came up with the idea of using a jet turbine to power an aircraft, and he began construction of a jet engine in 1935. It was tested in 1937, and an airplane using his engine first flew in 1941.

Like Whittle, Ohain was only twenty-two when he conceived the idea of a jet-propelled aircraft. His design was similar to Whittle's, but it differed in the internal arrangement of the parts. An airplane using his design for an engine was first flown in 1939. So both Germany and England actually had jet engines before the beginning of the war. But only Germany used the technology for a new type of fighter before the end of the war.

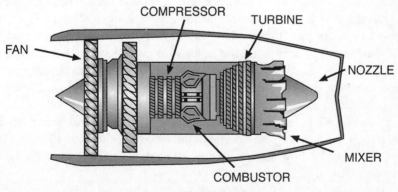

Details of a jet engine.

Jet engines operate as a result of Newton's third law, which states that for every action there is an equal and opposite reaction. The opposite reaction is what gives the thrust that pushes the jet plane forward. The easiest way to visualize this is to blow up a rubber balloon and let it go. You see immediately that it flies off in an array of flips and loops as the air forces its way out of the balloon. In short, as the air pushes its way out, it forces the deflating balloon in the opposite direction. This is basically what happens in a jet engine.

Several different kinds of jet engines now exist, but we'll restrict our discussion to the turbojet. At the front of the turbojet is an inlet that allows air to enter. Once inside, the air is compressed by blades that squeeze it into a much smaller volume. From here it is forced into what is called the combustion chamber. With the increase in pressure, the temperature of the gas goes up until it reaches over a thousand degrees Fahrenheit. Fuel is then sprayed into the air, and the mixture is ignited. This causes it to heat even more dramatically, and it leaves the combustion chamber, or combustor, with a temperature of about three thousand degrees Fahrenheit. The resulting heated gas exerts a large force in all directions, but it exits only at the rear of the engine, and this gives the plane a tremendous forward thrust. As the gas leaves the engine it passes through a series of blades that constitute the turbine, which rotates the turbine shaft. The turbine shaft, in turn, rotates a compressor that brings in a new supply of air. Thrust can be increased with the use of what is called an afterburner, where extra fuel is sprayed into the exiting gases, which burn to provide additional thrust.

THE FIRST ROCKETS IN WAR

Not only was the first jet introduced in World War II, but so was the first large ballistic rocket. Much of the technology, however, had already been developed by the physicist Robert Goddard. Goddard is now often referred to as the father of modern rocket propulsion, and the NASA Goddard space Center in Maryland is named after him. Most of his work took place at Clark University at Worcester, Massachusetts, where he was head of the physics department. In 1926 he constructed and launched the first rocket using liquid fuel. Earlier, in 1914, he had patented both liquid rocket fuel and solid rocket fuel. He made many contributions to rocketry, including a gyroscope control, power-driven fuel pumps, and vanes on the exterior of the rocket to help in its guidance. And he was the first to show that a rocket would work in vacuum and that it didn't need air to push against.

Early in World War II the Germans became interested in the possibility of using rockets as weapons. Artillery Captain Walter Dornberger was assigned the job of determining how effective they would be. While looking into the problem, a young engineer by the name of Wernher von Braun came to his attention, and he hired him as head of his rocket artillery unit. By 1934 von Braun had a team of eighty engineers working for him, and operations were moved to Peenemünde, on the Baltic coast. Hitler now began taking an interest in the project.

Wernher von Braun.

Von Braun and his team had many problems to overcome. Rockets look rather simple, but a lot of science, particularly physics, is needed to make them work properly. The V-2 that von Braun was building could reach an altitude of almost seventy miles, and at this altitude there is almost no air. And the rocket fuel needed an ample supply of oxygen for it to burn. This meant that oxygen had to be added to the propellant. The V-2 used a 75 percent ethanol-water mixture for fuel and liquid oxygen as an oxidizer.[12]

Rockets are propelled in the same way jets are propelled. They also work because of Newton's third law, and again it's the reactive force that produces the thrust. It's also important to note that the rocket flight consists of several phases: launch, thrust, cruise, and crash. Actually, the first phase (launch) is when the rocket is sitting on the launch pad, so it's not moving. At this point there are two forces acting on it: the weight of the rocket downward, and its reaction force acting back from the pad. These two forces are equal and opposite.

The thrust phase begins when the rocket engine begins firing. At this time there will be three forces acting on the rocket: the weight of the rocket, the thrust provided by the engine, and a drag force that is a result of air resistance. If we now apply Newton's second law, which states that force equals mass times acceleration, we get $F_{thrust} - F_{drag} -$ wt. $= ma$, where m is mass, a is acceleration, and wt. is the rocket's weight. There is a small problem here, however: the mass of the rocket changes as a rocket moves upward because fuel is being burned. But this was easily overcome by early engineers.

The blast from the engine will eventually stop at some point, and the rocket will enter the cruise phase. During this time there is no longer an upward thrust on the rocket, and it is on its own. It will continue gaining altitude for some time after its engines are shut down because of its velocity, but eventually it will reach its maximum altitude and begin falling back to earth, and because of gravity it will accelerate according to the formula $a = (wt. - F_{drag}) / m$. Thus, except for drag, it will drop like a falling stone. In reality, of course, the rocket is not going straight up and down, it is also moving horizontally, so its path will generally be similar to that of an artillery shell.

In a liquid-fueled rocket, the propellant and oxidizer have to be kept in separate tanks before the combustion. Oxygen is then combined with the fuel, with mixing taking place when the oxygen and fuel are sprayed into the combustion chamber. The ignition gases exit through a nozzle at the lower end, producing the thrust. These gases are at a very high temperature, so the nozzle has to be cooled. In early rockets the exhaust was cooled using alcohol and water.

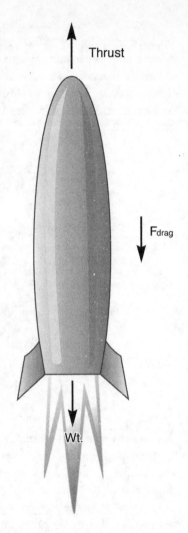

Rocket, showing thrust, drag, and weight.

The rocket also had to be stabilized once it was in flight, otherwise it would tumble uncontrollably. Two types of systems have been used for this: active and passive. Active elements are movable and passive are fixed. Of critical importance is the center of gravity of the rocket. It is important because all objects, including rockets, move around their center of gravity when they tumble. The center of gravity is the same as a center of mass, namely the point where all the mass can be considered to be concentrated.

In flight a rocket can tumble around one or more of three different axes, referred to as the roll, pitch, and yaw axes. Spin around the roll axis is no problem, but we want to avoid tumbling around either of the other axes. Gyroscopes are used for this, and also to assist in guidance. The vanes on the lower end of the rocket also help stabilize it.

The V-2 was to be Hitler's vengeance weapon, and in early September 1944 he declared that V-2 attacks would begin, and London was to be a major target. Over the next few months over fourteen hundred were directed at London. But their accuracy was poor and they were unable to hit vital targets. For the most part, the V-2 was a terror weapon, and it did, indeed, create a lot of terror as it shot across the English sky. Because of the speed of V-2 rockets (approximately 2,200 miles per hour) and their high-altitude flight, they were almost impossible to shoot down. In all, about 2,550 civilians were killed in London by V-2s, and another 6,500 were injured.

The Germans also built another, similar weapon called the V-1 "buzz bomb." It was smaller than the V-2, with a length of twenty-seven feet, compared with the V-2's forty-six feet, and it was much slower. A pulsed jet engine powered it; air entered the intake of the engine where it was mixed with fuel and ignited by spark plugs. Shutters opened and closed at the rear of the device about fifty times per second, giving it the buzzing sound that inspired its nickname.

The V-1 was developed at Peenemünde at the same time that the V-2 was being built. It was not a ballistic rocket; rather, it was launched from ground sites using a ramp and catapult. It is therefore referred to as a cruise missile. The first V-1 attack took place in mid-June 1944, just before the V-2 attacks began, and the V-1 attacks were also directed toward London. Like the V-2, the V-1 could not attack specific targets, so it, too, was mainly meant as a terror weapon. But unlike the V-2, there was considerable defense against it. Some of the faster airplanes could knock it down in flight, and it was quite vulnerable to coastal artillery. In fact, by late August 1944, almost 70 percent of incoming V-1s were being destroyed by coastal artillery. In all, about ten thousand V-1s were fired at England. About 2,420 reached London, killing approximately 6,180 people and injuring 17,780.

240   THE PHYSICS OF WAR

OTHER WEAPONS AND SMALL ARMS

Tanks played a large role in World War II. During the German blitzkrieg, in fact, they seemed to be unstoppable, and the Allies were soon looking for weapons that could counter them. Over the next few years several types of warheads were developed that were able to penetrate the armor of a tank, and they employed an important physics principle. They were based on the idea of a shaped charge. A shaped charge is an explosion that has a shape that focuses the energy of the shell. It is based on what is called the Munroe effect, discovered by the American chemist Charles Munroe. Munroe showed that a hollowed end on a charge produces a much more powerful wave that concentrates the explosion along the axis of the charge. This is because the shock waves from the explosion are reinforced in this case.

When applied to stopping tanks, the warheads are referred to as HEAT warheads (high- explosive, anti-tank warheads). They create a high-velocity stream of metal that can push through relatively heavy tank armor. This stream actually moves at nearly twenty-five times the speed of sound. HEAT warheads are less effective if they spin, so they usually are fin-stabilized.

HEAT rounds caused a significant change in tank warfare when they were first introduced late in the war. A single soldier could now destroy a tank using a hand-held weapon. The search was soon underway for a protection from the new shells, and the Germans began protecting their tanks with armored or mesh skirts, which caused the HEAT shells to detonate prematurely.

Another type of shell was also used quite effectively against tanks. It was called the HESH warhead (high-explosive squashed head). It was originally developed for penetrating concrete buildings, but it was also found to be effective against tanks. In this case the explosive material is "squashed" when it hits the target so that it spreads out over a large area. A detonating fuse triggers it at this point, creating a larger shockwave due to its larger area. This shockwave moves through the metal to the interior of the tank, causing pieces of metal to fly off the interior wall at high speed. These metal pieces could injure or kill the crew and ignite ammunition or fuel inside the tank.

Both HESH and HEAT warheads were delivered against armored vehicles using bazookas. A bazooka is a rocket-powered, recoilless weapon originally developed by Robert Goddard while he was working on rocket propulsion. He and a coworker Clarence Hickman developed and demonstrated it to the US Army at the Aberdeen Proving Ground in Maryland in November 1918. At this point, however, it didn't use a shaped charge. It was teamed up with shaped charges in 1942, and it was first used in North Africa and by the Russians on the Eastern front at about the same time. The early models were not too reliable, however, and some of them were captured by the Germans. The Germans quickly copied and improved on the early bazookas, and much to the surprise of the Allies, the German bazookas were more powerful than theirs and had greater armor penetration.

Another important development in which physics was involved was the proximity fuse. At the beginning of the war detonation of a warhead occurred when it hit the target, or after a certain time set on a timer. Both of these had disadvantages, and the full effect of most exploding shells was not realized. With the proximity fuse, the device detonates automatically when the distance between the target and the projectile is smaller than some predetermined value. Shells could therefore be made to detonate before they hit the ground—in particular, over the heads of enemy troops—which improved their effectiveness.

The fuse was based on electromagnetic principles; it contained an oscillator connected to an antenna that functioned as both the transmitter and receiver. As the shell closed in on the target it could determine how far it was away by analyzing the reflected signal. It was used quite effectively against V-1 buzz bomb attacks on England as well as during the Battle of the Bulge. It was also helpful in the defense against Japanese kamikaze attacks in the Pacific.

Radio-guided missiles were also used for the first time in World War II. The Germans developed an antiship guided bomb called the Fritz X. It was delivered by aircraft and was radio-controlled from the delivering plane. Signals were picked up by a receiver in the missile. Fritz X was not considered to be very successful, however. Similar guided bombs were also developed in England. Called GB-1s, they were dropped on Cologne, Germany. Another German guided bomb was the Kraus X-1; several Allied warships were heavily damaged by it. And the V-1 and V-2 were also radio guided.

Another of the ingenious devices to come out of the war was the Norden bombsight.[13] One of the major problems during the early part of the war was accurate bombing from high altitudes. In 1943 a plane dropping a bomb from a high altitude had a CEP (circular error probability) of twelve hundred feet, which made the likelihood of hitting a target extremely low. It was so low, in fact, that both the air force and the navy had given up on pinpoint bombing attacks. Over several years, however, Carl Norden, a Dutch engineer who had immigrated to the United States, had been working on a bombsight. One of the main problems in using bombsights was leveling the aircraft so that the sight could be pointed straight down. Wind was also a serious problem. Norden's bombsight allowed bombs to be dropped at exactly the right time for hitting a given target. It used an analog computer consisting of gyros, motors, gears, mirrors, levels, and a telescope. The bombardier would program the airspeed, wind speed, direction, and altitude into the device. The computer would then calculate the trajectory needed for the bomb to hit the target. Then, as the plane approached the target, the pilot would turn the plane over to autopilot so that it would fly to the precise point needed for the drop. It is said that with this device a bomb could be placed within a one-hundred-foot circle from a height of four miles.

The Norden bombsight was one of the major secrets of the war, and its existence was carefully guarded for the duration of the war. It was particularly effective in the bombing of Germany during the later parts of the war.

Finally, let's look at the small arms and infantry weapons that were used during the war. They were much more powerful, accurate, and lethal than those used in World War I. At the beginning of the war, however, some of the same weapons were used. The bolt-action rifles used in World War I were also used at the beginning of World War II. Later on they were used as sniper rifles, mostly because of their long range and high accuracy. A bolt-action rifle equipped with a telescopic sight was an excellent sniper weapon, but for close-up fighting soldiers needed a much faster rate of fire, and because of this, semiautomatic rifles were soon developed. One of the best American semiautomatics was the M1 Garand, and it soon became the standard American rifle of the war.

The submachine gun also played a large role in the war. It was the small, relatively light equivalent of the regular machine gun. Its ammunition, however, was much smaller and lighter, and this meant that it had a relatively short range, and its accuracy was not as high. But it was quite effective in short-range combat. The Germans used it extensively; their best-known submachine gun was the MP-18. The American equivalent was the Thompson submachine gun.

The major problem with the submachine gun was its inaccuracy and short range. In most battlefield situations soldiers needed both rapid fire and accuracy at a distance. The accuracy did not need to be as great as that of a standard bolt-action rifle, such as the Lee-Enfield or the Springfield, but a range greater than that of the submachine gun was desired. Because of this, the assault rifle was developed. It was first used by the German army; their MP-43 came into service in 1943 and was clearly a superior weapon. The American M-16 and Russian AK-47, which came into being after the war, were based on it.

Basic machine guns were still used, as they were in World War I, but they were now much lighter so that they could be handled by a single soldier. In most cases, however, a second soldier was needed for carrying ammunition and to help set it up and feed it during firing. Finally, other weapons such as hand grenades, flamethrowers, and light mortars of various types were also used. And most were more lethal because of technical advances.

COMPUTERS AND INTELLIGENCE

Another area in which tremendous advances were made as a result of the war was that of computers. World War I was perhaps the first war in which a large amount of information had to be moved as quickly as possible, and for this, a good communication system was needed. And of course, the need became even greater in World War II. Not only was there a need for communication about the movement and direction that various troops, squadrons, and so on should take, but it was also important to keep this information from the enemy. This meant that it had to be enciphered, which soon set off a race between code breakers and code makers. Codes became more and more complicated, and soon they could only be deciphered by machines, namely computers. Work on computers had begun before the war, much of it in Germany. The German engineer Konrad Zuse had built a simple computer that he called the Z1 in 1936. He continued to work on it during the war, improving it significantly. A similar device, eventually called Mark I was being built in the United States.[14] The war, and particularly a need for decoding enemy ciphers, soon created a demand for larger and faster computers. The Germans had begun using a coding machine called Enigma. Enigma allowed an operator to type a message then scramble it using notched wheels or rotors, each of which contained the letters of the alphabet. There were twenty-six electrical contacts on each side of the wheels corresponding to the letters of the alphabet. When a message was typed in, it was sent to the second wheel via electrical contacts, but contact was made at a different position on the second wheel, so a given letter, such as C, would be given a different designation, such as Z. Contact was then passed from this wheel to a third wheel, and again contact was made at a different position. In the earliest models, three wheels were used, but more wheels were added, making it even more complicated. With such a setup, it was almost impossible for someone to decode its messages. Furthermore, the codes could be changed each time the machine was used. Decoding was simple for the receiver, however; he merely had to set his machine up in the same way as the sender's machine.

Polish intelligence was the first to break the code; with the help of a German spy and using some complicated mathematics they managed to break the code in 1932, and they continued to decode German messages up to 1939. With the outbreak of the war, however, the Germans increased their security by making the system ten times more complicated. It was now beyond the Poles, so they handed everything they knew over to the British code breakers. The British code-breaking unit, codenamed Ultra, was set up at Bletchley Park in Sussex.

The British began working on the code, but they made little progress until Alan Turing joined the group. In addition to mathematics, he had studied cryptology at Princeton, where he had obtained his doctorate, so he was well-equipped to tackle the enigma code. He soon built a machine he called the bombe, which cracked the code. The bombe searched for possible "correct" settings of the Enigma that had sent the message. Billions of possible settings had to be searched, but his machine was fast (for this time), and it would eventually narrow in on the correct setting. But there was a problem: Turing and company were allowed to build only a few bombes, and large numbers were needed to decipher all the incoming German messages. Turing and his coworker Gordon Welchman were frustrated and didn't know what to do. Finally, against all rules, they wrote directly to Winston Churchill. Churchill replied immediately, giving priority to their request. Over the next few years over two hundred bombes went into operation.[15]

Alan Turing.

Enigma was used by the German navy, army, and air force, but the German high command used another, even more complicated encoder called Lorentz. It was introduced in 1941, and it used twelve wheels. The only way to break its code was with the use of a very large computer, much larger than anything that had ever been built. It would be a huge undertaking, but the information it could supply would be of tremendous value. The design engineer was Tommy Flowers, and the prototype, called Colossus Mark I, was produced in December 1943. It was in operation by February 1944.

With the Colossus, the messages sent by the Lorentz machine could be decoded, and over the next few months a large amount of German intelligence was intercepted and decoded. The Colossus, along with Turing's bombe, no doubt helped to shorten the war.

17

THE ATOMIC BOMB

We saw in the last chapter that physics played an important role in many of the weapons of World War II, but it played an even larger role in the greatest weapon of the war—the atomic bomb. Indeed, it played a central role, for the atomic bomb is possible only because of our knowledge of fundamental physical concepts in physics. The subatomic particles that constitute the nucleus within the nucleus are bound together by what is called binding energy, and it is this binding energy that makes the atomic bomb possible.

The development of the atomic bomb is, without a doubt, one of the most impressive and awe-inspiring developments in history. Not it only did take a number of fundamental breakthroughs by a few ingenious thinkers, but it also took a tremendous effort by thousands of people to achieve it. And not only did these people achieve a goal that seemed almost impossible to many at first, but it showed what could be accomplished with enough motivation, determination, and ingenuity.

THE BEGINNING

It's hard to say exactly how it all began, but the experiments of James Chadwick of Cambridge University in England were critical. He was repeating an experiment done earlier by Irene Joliot-Curie and her husband Frederic Joliot-Curie in which a strange particle was able to knock protons out of paraffin. The Joliot-Curies thought the strange particle was a gamma ray. Chadwick showed that it was actually a neutrally charged particle, which he called a neutron.[1] This neutral particle was in the nucleus of the atom along with the proton. As it turned out, the neutron had the same mass as the proton. The sum of the atomic weight of the protons and the atomic weight of the neutrons in an atom is approximately equal to the atomic weight (A) of the atom. We also designate the total number of protons in the nucleus as the atomic number (Z). And from these two numbers we can easily determine the number of neutrons in the nucleus; it is just A – Z. As an example, consider the hydrogen atom, which we know has a nucleus made up of a single proton. It has A = 1 and Z = 1, and since A – Z = 0, it has no neutrons. In the same way, the helium atom has A = 4 and two protons, or Z = 2, and since A – Z = 2, it has 2 neutrons. We can continue in his way through all the elements.

The neutron turned out to be a particularly important research particle because it was neutral. Early physicists tried to learn more about the nucleus by projecting high-speed particles at it to see what would happen. The only known particles available at the time, however, were the proton and the electron, but the electron was too light to have any effect on the nucleus, and the proton was positively charged, as was the nucleus, so the nucleus and the proton repelled one another. Because of this, the proton was also an ineffective projectile. The neutron, however, was not repelled electrically by the electrons or nucleus, so it was an ideal projectile. Before we look into how it was used, however, let's consider Einstein's contribution to the atomic bomb.

EINSTEIN'S ROLE

Einstein is sometimes called the father of the atomic bomb, a title he abhorred, and in reality he had very little to do with it directly. But he did make an important contribution. In a short paper published shortly after he published his famous paper on special relativity in 1905, he showed that energy and mass were related. The title of the paper was "Does the Inertia of a Body Depend on Its Energy Content?" It was only three pages long, but it was one of the most important papers ever published. This paper, along with a paper published a year later, showed an equivalence between mass and energy. In particular, it gave us the equation $E = mc^2$, where the energy (E) of a given amount of mass is equal to the mass (m) multiplied by the speed of light (c) squared. The speed of light is 186,000 miles per second, and if you square it (multiply it by itself), you obviously get a very large number. This tells us that there is a large amount of energy associated with even a very small amount of mass. Fortunately, it is very difficult to transfer mass directly into energy, but this is, indeed, what happens in an atomic explosion.[2]

THE ITALIAN BREAKTHROUGH

Most physicists are either experimentalists or theoreticians. Enrico Fermi of the University of Rome, however, was one of very few people who excelled in both areas. He made important theoretical contributions, but at the same time he was a first-rate experimentalist. When the neutron was discovered in 1932, Fermi immediately realized that it would make an ideal projectile. It would not be repelled by the nucleus, and it could easily be projected fast enough so that the surrounding electrons would have no effect on it. The problem was finding a good source of neutrons, and he was soon able to devise an apparatus that would produce a beam of neutrons.[3]

Enrico Fermi.

One of the hottest areas in physics at the time was radioactive decay. A number of elements were known to spontaneously decay, emitting various types of radiation, referred to as alpha, beta, and gamma rays. A number of people, including Marie Curie, had made important contributions to the area. In 1934, however, Irene Curie and Frederic Joliot announced that they had been able to induce artificial radioactivity. In other words, they had caused a stable element to become radioactive. They had bombarded aluminum nuclei with alpha particles and caused it to become radioactive. They had also found that boron responded in the same way when bombarded with alpha rays.

Fermi was fascinated by the result, and he was sure he could improve on it.[4] The alpha particles were big and heavy, and they could easily be stopped, even by a sheet of paper. Furthermore, they were charged. Neutrons would make a much better projectile, and he now had an apparatus for producing them. In addition, he had improved a device that had been invented several years earlier, called the Geiger Counter, which was used for measuring the radiation produced. Fermi and his group began by repeating the experiments of Joliot and Curie, and they quickly verified their results. They then turned to heavier elements. And indeed, many of them became radioactive, but in most cases the radioactivity was short-lived. Some of the elements, in fact, had half-lives (half-life is the time it takes for a mass of radioactive material to fall to half its original value) of less than a minute.[5]

Fermi and his team tested most of the elements of the periodic table in this way, all the way up to the heaviest known element at the time, namely uranium. And uranium was of particular interest to him. There were no known elements heavier than uranium, so he wondered what would happen if he shot a neutron at the uranium nucleus and it was absorbed. Would a new element be formed? Uranium has an atomic weight of 238 (total number of protons and neutrons in the nucleus); if it absorbed a neutron it should become uranium-239. But this created a new problem: how could they detect uranium-239? This proved to be frustratingly difficult. Finally, however, the team identified a slightly heavier element. Fermi was overjoyed. He had created an element beyond uranium-238. With this he concluded his experiments, but in doing so he failed to make one of the greatest discoveries in history.

In the meantime the world around him was becoming more and more turbulent. Hitler had seized power in Germany and Mussolini had signed a pact with him. Hitler's war against the Jews had already begun, and he demanded that Mussolini cooperate. Jews in Italy were therefore subjected to new legal restrictions. Fermi himself was not in danger, but Laura, his wife, was Jewish, and Fermi knew she might eventually be rounded up by authorities. He wasn't sure what to do. Government officials were unlikely to allow him to leave the country with his wife. Earlier he had been offered several positions at universities in the United States, but he had turned the offers down. He decided to write and ask if they were still interested, and indeed he got an offer from Columbia University. The problem now, however, was getting out of the country without arousing suspicion.

The breakthrough he was waiting for came in the fall of 1938. Fermi was in Copenhagen for a physics meeting when Niels Bohr took him aside and told him he was in line for the Nobel Prize that would be awarded later that year. Fermi was excited, not only at the prospect of winning the prize, but also because it might provide a possible route out of Italy. Indeed, a few weeks later he received a call informing him that he had won the prize, and that he would have to go to Stockholm, Sweden, to collect it. Furthermore, he was invited to take his family with him to Sweden. Immediately after the Nobel-Prize ceremony, Fermi boarded a plane for England, accompanied by his wife and children. From there they boarded a ship to New York.

HAHN, MEITNER, AND STRASSMANN

Lise Meitner was born into a Jewish family in Vienna, Austria, in 1878. She became interested in physics at an early age, but a scientific vocation was difficult for a woman at that time. Nevertheless, she managed to get a doctoral degree in physics at the University of Vienna. After obtaining it she went to the Kaiser Wilhelm Institute in Berlin and began working as an assistant to the chemist Otto Hahn.[6] Initially she worked with no salary, but eventually she became head of a section in chemistry. She worked with Hahn for thirty years, making several important discoveries.[7]

When Hitler came to power in 1933 she was acting director of the Institute of Chemistry. Although she'd been born into a Jewish family, she had converted to Christianity early in her life, and as an adult she identified herself as Lutheran. Furthermore, she was Austrian by birth. So initially she wasn't worried by Hitler's action against Jews. She buried herself in her work. Others, including the Joliet-Curies in Paris had followed Fermi's lead in bombarding heavy elements, particularly uranium, with neutrons. Hahn and Meitner soon took an interest in their work.

Meitner had barely begun the work, however, when Hitler annexed Austria and issued a proclamation against all Jews, including those from Austria. Although she no longer considered herself a Jew, Meitner knew that that didn't matter to the Nazis. She had to get beyond the German area of influence as soon as possible, but there was a problem. Her visa had expired, and she could not apply for a new one because it would alert the authorities. She was uncertain about what to do, so she wrote Niels Bohr in Copenhagen. He made arrangements for her to get to Holland without a visa. But she still had to get past the Nazi patrols at the border. And, as she feared, a Nazi officer at the border asked her for her visa. She knew it had expired, but she handed it to him. He looked it over carefully as she sat in a state of fright. Finally, after several minutes he handed it back to her without saying anything. Minutes later, much to her relief, she was in Holland.

Bohr got her a position in Stockholm, Sweden, but it came with almost no support, and she was soon quite unhappy. Furthermore, Hahn and his assistant, Fritz Strassmann, were continuing the experiments they had started. Fermi had assumed that when uranium was bombarded with a neutron it would create a heavier, transuranic element, but he hadn't proved it beyond a doubt. But when Hahn and Strassmann did the experiment they were thoroughly confused. They couldn't verify Fermi's result; furthermore, an element, namely barium, with only about half the atomic weight of uranium, appeared to have been produced. It didn't make any sense, but Hahn had run the experiment through several times, getting the same result each time. Knowing that Meitner had a much better knowledge of nuclear physics than he did, he sent her a letter asking her if she had an explanation.

CHRISTMAS 1938

Meitner was amazed by the result, and confused. She had no explanation, but she was sure that Hahn had not made a mistake. If he said there was barium present after the bombardment, it had to be true. But where did it come from? Christmas neared, she pondered the strange result. She had a nephew, Otto Frisch, who was working for Bohr in Copenhagen, and she knew he was single. So she wrote to him to ask whether he would like to spend Christmas with her. He wrote back saying he would be delighted to spend it with her. He had been working on an interesting project related to the magnetic properties of the nucleus, and he was anxious to tell her about it, as she might have some suggestions for him.[8]

He was a little disappointed, however, when he met her. She immediately began talking about the letter she had received from Hahn. She finally handed it to him to read. Frisch suggested that the strange outcome of Hahn's experiment might be an error due to contamination, but she argued that Hahn was too good a chemist to allow that. They continued talking about it for some time.

In general, only small particles such as electrons, neutrons, and alpha particles were observed in nuclear reactions. A heavier or slightly lighter nucleus might be produced, but there seemed to be no way that a nucleus with half the atomic weight of uranium would be produced. The only way it could be produced is if the uranium nucleus had somehow broken in half. But that was impossible—the energy required for something like that had to be incredibly large, and the neutrons that hit the nucleus only had a small energy.

Frisch had brought his skis and wanted to do some cross-country skiing while he was there, so off they went, Frisch on skis and Meitner walking in the snow. They began talking about the two models of the nucleus. Ernest Rutherford had suggested it was a small, rigid ball, but Bohr had recently put forward a new and different model that was quite controversial at the time. He suggested that the nucleus was actually relatively soft and pliable—more like a drop of water.

There was no way Rutherford's model would allow a splitting into two nuclei, each half the size of the original one. Bohr's model, however, might work. They paused and sat down on a fallen tree near the path. Lise pulled out a piece of paper and a pencil from her pocket. She drew a picture of a uranium nucleus, assuming that it was a sphere. What would happen to it if a neutron hit it? If it was like a drop of water, it could change its shape slightly; it could become elongated. Meitner began to calculate the forces on the drop. Cohesive forces held the drop together, so if it eventually broke apart, these forces would have to be overcome. And the cohesive force was related to the surface tension of the drop. The force that might over-come it had to come from the charge of the nucleus. And indeed, the large, unstable uranium nucleus would likely wobble. If so, it would become elongated at first, but as it continued to oscillate it might begin to resemble a dumbbell, and if this happened, the two masses at the ends of the dumb-bell would repel one another as a result of their similar charges.[9]

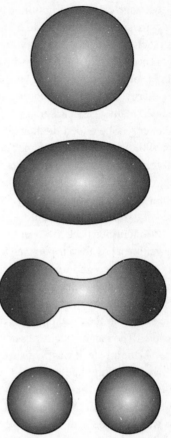

A wobbling drop that fissions into two smaller drops.

Meitner calculated how much energy would be released if this occurred. She was surprised to find that it would be about 200 million electron volts (an electron volt is the energy an electron gains in passing through a voltage difference of 1 volt). This was not a large amount, but when multiplied by the number of nuclei that would be splitting, it would be very large. But where did this energy come from? Meitner immediately thought about a lecture she had attended many years earlier at which Einstein had given a formula relating mass and energy. She added the masses of the two product nuclei and compared the sum with the mass of uranium. Then she used Einstein's formula to convert the difference in mass to energy. Amazingly, the result was the same: 200 million electron volts. This was obviously not a coincidence. Uranium nuclei had split in half—an amazing discovery if indeed that was what had happened. They decided to publish their results as soon as possible.

Frisch rushed back to Copenhagen. He could hardly wait to tell Bohr. But Bohr was getting ready for a trip to the United States and couldn't spend much time with him. Nevertheless, he was delighted with the news, and he encouraged Frisch and Meitner to publish as soon as possible. Frisch began writing up the paper, but he was stumped by the problem of how to describe the splitting. A friend noted that it was quite similar to the breaking apart of a simple cell in biology, and that was called "fission." The name "nuclear fission" immediately came to mind, and he used the phrase in the article. It was published five weeks later in the scientific journal *Nature*.

By then Hahn had published his result, but Meitner had not yet told him of the interpretation that she and Frisch had developed, so there was no mention of fission in his paper. Meitner, in fact, hesitated for a while before she told Hahn. She wanted to be sure that the paper she and Frisch had written was published first. There was some irony in all of this, however. Hahn was awarded the Nobel Prize in 1944 for his discovery of fission, with no mention of Meitner, even though she was the one who interpreted his result as fission.

A CHAIN REACTION

Bohr could hardly contain his excitement about the new discovery as he sailed to America. Along with coworker Leon Rosen, he tried to work out the details of what might happen during the fission of a uranium nucleus. There was no doubt that a tremendous amount of energy would be released. Could it be used to make a bomb? The possibility worried him. He had promised Frisch that he wouldn't mention the discovery until after Frisch and Meitner had published their results, but he forgot to mention this to Rosen.

Bohr, Rosen, and their group were met in New York by Fermi, Fermi's wife, and John Wheeler, a former student of Bohr's. Bohr said nothing about the discovery, but within a short time he discovered that everyone seemed to know about it. Then he realized he had forgotten to tell Rosen to keep it secret. The secret was now out, so he decided to make an announcement at a Washington conference on theoretical physics that he would be attending within a few days. Many of the world's top physicists were at the meeting, including Hans Bethe, Edward Teller, George Gamow, Harold Urey, Isidor Isaac Rabi, Otto Stern, and Gregory Breit. As expected, everyone was stunned when the news was announced, particularly after Bohr mentioned that a super bomb might be possible using nuclear fission.

When Fermi heard about the discovery, he had mixed emotions; he realized he had come very close to making the discovery himself, and he was annoyed. But at the same time he realized it was a momentous discovery, and it was important to follow up on it as quickly as possible. He immediately set up a simple experiment at Columbia University to verify the result, and he was pleased to see that there was no doubt: uranium nuclei did, indeed, fission.

Everyone was talking about the new discovery, and as Bohr, Wheeler, Fermi, and Leo Szilard got together for dinner the day after the conference they were still tossing around ideas about it. One of the most interesting of these ideas was one that Bohr had casually mentioned at the meeting. He pointed out that if the uranium nucleus split in half, leaving two lighter nuclei, there would have to be some neutrons left over, but he wasn't sure how many. Nevertheless, if there were two or more neutrons coming out of the reaction, each of them might produce a new fission. Furthermore, it was like the old story of the employer offering a new employee a wage of one cent the first day, then doubling the wage each day thereafter. After rejecting it, the employee realizes that he would have been a millionaire within a month. In essence, it doesn't take much doubling before small numbers become huge. And since each of the new fissions would take place in a tiny fraction of a second, an incredible amount of energy would be released very rapidly.[10]

The possibility was so exciting that Bohr asked Wheeler if he would like to work in collaboration with him to see what was possible, and Wheeler agreed. But they soon found that they would need some additional experimental results, so an experiment was set up at Princeton University to find out how the rate of fission would be affected by the speed or energy of the incoming, or bombarding, neutrons. In particular, they wanted to find out if there was a significant difference between slow and fast neutrons. They began bombarding uranium with extremely energetic neutrons, and, as expected, the higher the energy of the neutron, the greater the fission rate. But they also got an unexpected result: at very low neutron energies, the rate of fission also increased. In essence, the fission rate was high for slow neutrons and also for very fast neutrons. This seemed a little crazy. Bohr and Wheeler thought about it. The reason for this had to be related to the uranium they were using, which was natural uranium that had come out of the ground.

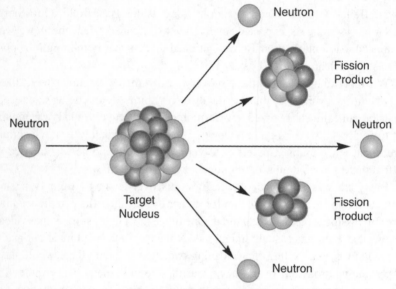

Fission creating a chain reaction.

To understand why this is important we have to go back to the elements and look more closely at how they are made up. As we saw earlier, they have a certain number of protons and neutrons in their nucleus (we will ignore the electrons because they are irrelevant for this discussion). Furthermore, each element is identified by a mass number (A) (closely related to the atomic weight), and an atomic number (Z). The mass number is equal to the number of protons plus the number of neutrons, while Z is the number of protons in the nucleus, and it's the number of protons in the nucleus that uniquely defines an element. For example, the carbon nucleus has six protons, but it can also have either seven or eight neutrons. This difference in the number of neutrons does not change it into a new element; rather, the different numbers of neutrons identify different isotopes of the same element. And, as it turns out, uranium also has two isotopes that differ in the number of neutrons they contain. They are referred to as U-238 and U-235. Natural uranium is a mixture of these two isotopes.

As Bohr and Wheeler looked closely at the results of the Princeton experiment they realized that the sudden increase in fission as a result of the bombardment with slow neutrons was due to U-235. The increase with fast neutrons was mainly due to U-238. This meant that U-235 fission required less energy than U-238 fission. Thus, U-235 would be much better for a bomb, particularly because secondary neutrons were very slow. The problem was that natural uranium consisted almost entirely of U-238; only 0.7 percent of natural uranium was U-235. And to make things even worse, because they were chemically the same, there was no chemical process that could separate U-235 from U-238. Some sort of physical process, such as diffusion, would have to be used, and it would be difficult to do.

Although they didn't realize it at the time, others were already thinking along the same line. Irene Joliet-Curie and her husband Frederic, in Paris, also realized that a bomb might be possible. Furthermore, Otto Hahn, who was still in Nazi-controlled Germany, would no doubt soon come to the same conclusion. In addition, Werner Heisenberg, one of the brightest and most famous physicists in the world, was also in Germany, along with several other world-renowned physicists. One who was particularly worried about the bomb-making implication of nuclear fission was Leo Szilard.

THE LETTER TO ROOSEVELT

Leo Szilard had come to America several years earlier from Germany. He was Jewish, and when Hitler came to power he knew his days in Germany were numbered. In 1933 he went to England, and later moved on to the United States. Interestingly, about this time he had already begun to think about the possibility of a super bomb. He told Fermi about his worry, but Fermi didn't take him seriously; at this stage Fermi was not yet convinced that a bomb could be built. Disappointed in Fermi's response, Szilard decided to do something about it on his own.[11] He knew that one of the largest known deposits of uranium in the world was in the Belgian Congo. And as soon as German scientists realized how important uranium was, they would rush to buy up as much of it as possible. Szilard had to stop them. He remembered that Einstein was a personal friend of Belgium's queen. He immediately phoned Einstein, who was now at Princeton's Institute for Advanced Study, but he was told that Einstein was at his summer home on Long Island.

Szilard acquired Einstein's address, but he now had a problem. He had never learned to drive a car, so he had to get his friend Isidor Isaac Rabi to drive him. After some trouble, they finally found their destination and were greeted by Einstein. Szilard told him the news, and Einstein was surprised; he had heard nothing about the new discoveries, but he was immediately concerned. He knew that if the Germans produced such a bomb they would likely use it, and it worried him. Szilard told him about the uranium deposits in the Belgian Congo and suggested that he write a letter to Elizabeth, the queen of Belgium. Einstein was reluctant to bother her, but he offered to write a letter to a friend who was in the Belgian cabinet.

They talked about other things they could do. A letter to the White House was suggested. Szilard knew, however, that a letter signed by him would be ignored whereas a letter from Einstein would be taken seriously. Einstein agreed to sign a letter written by Szilard. The next problem was getting the letter to Roosevelt; it had to be delivered to him directly to have any impact. Szilard remembered an acquaintance by the name of Alexander Sachs, who sometimes visited Roosevelt. Szilard gave him the letter on August 15, 1939, and Sachs agreed to deliver it.

Germany was on the verge of attacking Poland at this time, however, and Roosevelt was particularly busy. Sachs tried several times but was unable get an appointment. He finally succeeded in October 1939. Roosevelt agreed that action was needed, and he authorized the creation of an advisory committee on uranium. The committee had its first meeting on October 21, at which six thousand dollars was budgeted for conducting experiments on neutrons. Szilard was disappointed with the small amount of money, but at least it was a first step.

Several problems would have to be overcome, however, before they could build a bomb. First of all, the uranium would have to be purified. At that time uranium had no known uses and very little of it had been mined, and the small amount that had been produced was not pure. Furthermore, Bohr and Fermi had showed that it was U-235 that was of most interest, and there was only a small amount of it in natural uranium. The U-235 would have to be separated out. And perhaps of most interest, some sort of device would have to be built to slow down the fission process so that it could be controlled. Such a device would be needed in order to determine whether a bomb would be possible. This device was eventually called a nuclear reactor. And for a controlled reaction, a moderator would be needed (a material that would absorb some of the neutrons coming out of the reaction). Two moderators were known: heavy water and graphite. Heavy water was expensive, so graphite seemed to be the better choice.

THE WAR BEGINS

In September 1939 Germany invaded Poland and World War II began. Because of Hitler's policies regarding Jews, many Jews had fled Germany, including Albert Einstein and other top physicists. But Werner Heisenberg, who had just won a Nobel Prize, was not Jewish, and he had no interest in leaving. He had been offered positions at several major universities in the United States, but he had turned them down. As a German, he felt compelled to serve his country in its time of need. And indeed, by the time the war started, the Nazi government had also heard about the possibility of a super bomb. In fact, a group of Germany's leading scientist (the few that were left) had been recruited to study uranium fission. The group was referred to as the *Uranverein* (Uranium Society), and one of the most prominent members was Otto Hahn, the man who had discovered fission.[12] Members of the Uranverein were reluctant at first to invite Heisenberg because he was a theorist, and the building of a bomb needed experimentalists. Furthermore, he had been friends with many Jewish scientists, including Einstein, so he was not considered desirable. Interestingly, Hahn was a rather reluctant member of the group, and in the end he made almost no contribution to it. He continued to raise objections against the project, sure that it would never be possible. The Uranverein finally decided to invite Heisenberg into the group in late September 1939 to help solve what seemed to be insurmountable problems. Soon he was leader of the group.

Heisenberg realized that the first step had to be the building of a nuclear reactor: a slowed-down bomb. And the best moderator was heavy water, or deuterium. But deuterium was not plentiful in Germany; however, it was being produced in a plant in Vemork, Norway. Germany had not yet invaded Norway, so would have to purchase the deuterium it needed. The Vemork plant was perched high above the fjords in a remote region of Norway, about 150 miles from Oslo.

The Germans approached the owner of the Vemork plant and offered to buy all its available heavy water. The Norwegians were surprised by the offer and wondered why the Germans would need so much. When they were not given an answer they refused to sell. About the same time, the Joliet-Curies in Paris had arrived at the same conclusion as the Americans and Germans. For a bomb, a reactor would be needed, and it would need heavy water as a moderator. The French government therefore also sent a representative to Vemork, and when he told Norwegian officials what the heavy water was needed for they promised him all they had at no cost.

Then, in April 1940, Germany invaded Norway, changing the situation. The German army raided the plant immediately and was disappointed to find that all the heavy water had been shipped to France. So the Germans immediately ordered an acceleration in production and insisted that everything that was produced was to be shipped to Berlin.

In early June 1940, Germany also invaded France, and when Paris fell on June 14, the Uranverein physicists immediately went to the Joliet-Curie laboratory, expecting to find heavy water and uranium. Joliet-Curie claimed that it had been loaded onto a ship that had been sunk, and the Germans accepted the answer. The materials had actually been shipped to England.

By now the Germans had made considerable progress. One of the Uranverein group had calculated that uranium would have to be enriched so that it contained 70 percent more U-235 than U-238 before it could be used in a reaction. They had also discovered that when U-238 captured a neutron it formed U-239, which was unstable and radioactively decayed in twenty-three minutes. The new element, which was unnamed, might be fissionable according to Carl Friedrich von Weizsäcker, one of the group members. This meant that if they could build a reactor they might have a way of producing another element that could be used for a bomb.

In the summer of 1940 a new building at the Kaiser Wilhelm Institute in Berlin was designated for the exclusive use of the Uranverein. It was next door to the Institute of Physics and was called the Virus House. Germany now had a good supply of uranium, ample heavy water, and several important developments from the Joliot-Curie lab. But the members of the Uranverein were soon stumped by the difficulty of separating U-235 from natural uranium.

MEANWHILE IN ENGLAND

In the meantime, Otto Frisch had not sat still; he had immigrated to England and was now working with Rudolf Peierls at the University of Birmingham. Peierls had previously worked for Fermi at the University of Rome. Frisch and Peierls collaborated on the problem of how much U-235 would be needed for a bomb. This was later referred to as the critical size. To their surprise, their calculations indicated that a very large amount was needed. Despite the setback, their work showed that a bomb was indeed possible. In fact, they concluded in their report that what was needed was two pieces of U-235 that were smaller than the critical size. When the two pieces were brought together, they would immediately explode, but they could be handled safely as long as they were below the critical size. Because of their work, British officials decided to set up an organization called the MAUD Committee for atomic research in early 1941.

No one was quite sure where the name came from. It is not an acronym, and it appeared to have come from a letter sent by Meitner to an English friend. She ended the letter with the name Maud, and for a while people thought it was some sort of code. It turned out that it wasn't.

In July 1941 the MAUD Committee issued two reports.[13] The first stated that it had been determined that a bomb could now be built using approximately twenty-five pounds of enriched uranium, and it would have the destructive effect of eighteen hundred tons of TNT. It recommended that work begin immediately and that it should be conducted in collaboration with the United States. At the time the United States had many more resources than England for building such a bomb. In addition, a new committee codenamed Tube Alloys was set up in conjunction with Canada for further developments of nuclear weapons.

In June 1940 the MAUD report was sent to Vannevar Bush, the head of the National Defense Research Committee in the United States. The results were reported to Roosevelt, and there was a considerable amount of discussion about the possibility of a bomb, but little action.

Finally, in August 1941, Mark Oliphant, one of the leaders of the MAUD Committee, decided to fly to United States to see what the problem was. To his dismay, he found that Bush had done very little with the report; it was sitting in a safe. He immediately met with several members of the American Uranium Committee and stressed the importance of action. He met with Ernest Lawrence on September 21, and they were soon joined by Robert Oppenheimer, who was surprised by what the British had achieved in relation to the bomb.

Both Oppenheimer and Lawrence were now committed. They contacted Arthur Compton of the University of Chicago, and a review committee was immediately set up. There was still some skepticism about U-235, but in the meantime Glenn Seaborg of the University of California had managed to create element 94 (now called plutonium). And on May 18 they showed that it had a rate of fission nearly twice that of U-235. It was therefore also a suitable material for an atomic bomb. This was good news. If a reactor could be built, this new element could be produced relatively easily.

In the meantime the MAUD Committee had sent further reports to United States. The second and third reports came in late October 1941, and they were much more urgent. They reported that a mass of about twelve kilograms was all that was needed for criticality.

Then, on December 7, 1941, planes from several Japanese aircraft carriers bombed the American fleet at Pearl Harbor. The next day Roosevelt declared war on Japan, and soon the United States was at war with Germany and Italy as well.

HEISENBERG AND BOHR

By December 1940, Heisenberg and his team had built their first reactor. It was a relatively simple device that failed to initiate a chain reaction. But other similar experiments were going on in Heidelberg and Leipzig. The one in Leipzig used paraffin wax as a moderator, and the one in Heidelberg used heavy water. Both experiments again failed. The problem appeared to be due to the uranium; it was not enriched enough with U-235. By now the Germans knew that if element 94 could be produced in a reactor, it would also make an excellent material for a bomb. So there was even more incentive to accelerate the project, and for this they needed more heavy water. Furthermore, they were now beginning to worry about the progress being made in England and the United States. Heisenberg was certain that his group was far ahead of them, but he had to know for sure.[14]

Denmark had been occupied by Germany in April 1940, but Bohr had decided to remain at his Institute for Theoretical Physics in Copenhagen. He was of Jewish descent, but the Danish government had, as part of its surrender, demanded that the Jews of Denmark be unharmed. Bohr would know what was going on in relation to the English and American projects, but how could Heisenberg talk to him? Both men were under close scrutiny.

When the Germans took over Denmark they set up a German cultural institute in Copenhagen. A proposal was given to the German foreign office to set up a symposium on theoretical physics and invite both Heisenberg and Bohr. It was set up for mid-September 1941. Heisenberg was anxious to talk to Bohr, but he worried about what his reaction might be. He had worked with Bohr and been on friendly terms with him for years, but now things were quite different. For his part, Bohr was not anxious to attend the conference, and he boycotted most of it. He was curious, however, about how much progress the Germans had made on atomic-bomb development, and he knew Heisenberg was part of the program.

Heisenberg visited Bohr's institute and had lunch with him twice. Bohr was quickly turned off, however, by Heisenberg. He told Bohr that it was critical that Germany win the war, as it would be helpful for the development of Eastern Europe. Bohr could not agree less. In their second meeting Bohr asked about the German atomic bomb and what progress had been made. Heisenberg knew the Gestapo, the secret police of Nazi Germany, was watching his every move, so he suggested they move to Bohr's study.

It soon became clear to Bohr that Heisenberg was doing everything possible to help Germany develop a bomb. This was a shock to Bohr. Heisenberg proceeded to make a sketch for Bohr. Bohr thought it was a drawing of an atomic bomb, and he was repulsed. As it turned out, it was actually a sketch of a reactor. Then Heisenberg started to probe him about the progress the English and Americans had made. Bohr was immediately suspicious that Heisenberg was trying to probe for secrets. Bohr said little, and the meeting, for the most part, was a disaster for both men.

THE MANHATTAN PROJECT

On October 9, 1941, Roosevelt gave the go-ahead for the development of the atomic bomb, and on December 6, the day before the bombing of Pearl Harbor, he authorized what would eventually become known as the Manhattan Project. Various projects were set up in laboratories across the United States, but there was little coordination, and many of the people involved began to get frustrated with the progress. Something had to be done.

Vannevar Bush, the director of the Office of Scientific Research and Development, suggested that a single person should be in charge and that the entire project should be directed by the Army Corps of Engineers. For many of the scientists this was not good news. They didn't like the idea of being bossed around by army officials. Bush selected Leslie Groves, a no-nonsense colonel with considerable experience in directing military construction projects, as military director of the Manhattan Project. Groves was not happy to take on the project at first.[15] He knew little about physics and didn't have much faith that such a bomb would work. Furthermore, he wasn't popular with the men who had worked under him because of his gruff approach, but everyone admitted that he got things done. And, as it turned out, he was the ideal man for the job.

Within a few weeks of taking his new assignment, at which point he'd been promoted to the rank of brigadier general, Groves went on an inspection tour of the various facilities throughout the country that were involved in the project. He traveled to Columbia University and talked to Harold Urey, who was involved with in the effort to separate U-235 from natural uranium, then he went to the University of Chicago to meet with Fermi. Fermi was now involved in building the first reactor. Then he went to the University of California at Berkeley, where he met Lawrence, who was then building a large particle accelerator called a cyclotron. He was impressed with some of what he saw but disheartened by other parts of the project. The basic problem seemed to be that there was no real organization and cooperation, and no sense of urgency.

At Berkeley he talked to Robert Oppenheimer, and almost from the moment he met him, he was impressed. Oppenheimer had a good overall grasp of what was needed to achieve the goal of producing a bomb, and he had considerable confidence that it could be done. His enthusiasm and confidence appealed to Groves. Groves had originally thought that he would select Lawrence as the scientific director of the project, but after meeting Oppenheimer he changed his mind. He was now sure that Oppenheimer was his man, and he suggested him at the next meeting of the Military Policy Committee.[16] After a few checks on him, however, it became obvious that there were problems. Everyone agreed that he was a first-rate scientist, but he had no experience in directing people. In addition, a check by the Federal Bureau of Investigation indicated that he might be a security risk, as some of his closest friends, including his brother, had been associated with the Communist Party. The FBI told Groves to find someone else. Groves went back and checked on other possibilities, but he came back even more convinced that Oppenheimer was the best man for the job. Stubbornly, he resubmitted Oppenheimer's name, and after some arguments, Oppenheimer was finally accepted.

Groves began conferring with Oppenheimer on how to approach the problem. Oppenheimer suggested that all the scientists should be brought together in a single lab or complex, and, as it turned out, this was what Groves had also been thinking. They would need a place that was relatively isolated so that the facility would not draw a lot of attention. Oppenheimer had spent much of his earlier life in northern New Mexico, and he thought it had all the qualifications needed. He remembered a place near Jemez Springs, about thirty miles north of Santa Fe. He had recovered from tuberculosis there in the summer of 1928. It seemed like an ideal place. A school called the Los Alamos Ranch School had been established there but was now on the verge of bankruptcy. Groves visited the area and agreed with Oppenheimer. He purchased the school and surrounding area immediately.

The project did not get off on a good track at first, however. Oppenheimer had thought that a group of about thirty scientists would be enough, and he was sure that there would be no problem directing them. Almost immediately Oppenheimer and Lawrence began scouring the country for the best scientists to bring to the new site. Some were reluctant to go, not sure if they would like the remoteness, isolation, and secrecy that would be required. Furthermore, the army was running things, in particular, the construction of the town at the site and the laboratories that would be needed. Something else that was bothersome to them was that Groves wanted compartmentalization. In short, he wanted each group to know everything about the particular aspect of the bomb they were working on but little or nothing about what other groups were doing. The fewer the people that knew everything about the project, the better. And secrecy had to be of the highest order; there would be no publication of any discoveries. Scientists did not normally work this way.

All of this became a problem for Oppenheimer, and his original group of thirty scientists grew to one hundred and then to fifteen hundred. For the first few months the place was a mess. Buildings, laboratories, roads, and many other facilities were being built, and with the spring thaw there was mud everywhere. It was his job to keep everybody happy. He had recruited some of the best physicist in the world, including Edward Teller, Hans Bethe, Felix Bloch, Richard Feynman, and Robert Serber, and he had lined up Enrico Fermi and Isidor Isaac Rabi as consultants (they were already involved in important war projects).

The problem before them seemed straightforward enough: get enough enriched uranium (high in U-235) to create a critical mass, and at the proper time bring two sub-critical masses together to create a chain reaction. The first calculations of the critical mass needed were not encouraging; it appeared to be very large, perhaps too large to be carried in an airplane as a bomb. But innovations were made so that the neutrons created in the blast were reflected back into the blast by a shield. This reduced the critical mass to about thirty-three pounds of U-235. At the same time it had now been shown that a bomb could also be made from plutonium, and only eleven pounds of plutonium would be needed. Of course, a reactor would have to be built to get the plutonium, so most of the interest was still in U-235.

In reality a slightly greater amount than the critical mass would be needed because of various problems; it was usually referred to as the super-critical mass. When two sub-super-critical masses were brought together they would create an explosive force equivalent to twenty thousand tons of TNT. But there was a serious problem: the masses had to be brought together very rapidly. If they came together too slowly, some of the mass would fission and detonate, which would blow other parts of the mass apart before they could fission. Calculations showed that they had to be brought together at a speed of 3,300 feet per second. However, this was beyond the highest speed produced by any explosive technique; the highest artillery speed known at the time was about 3,100 feet per second.

Furthermore, there was another problem, and it was associated with the neutrons that would be triggering the fission. All that was needed was one neutron to start the chain reaction, but it had to be delivered exactly when the two halves came together. The problem was that there were neutrons all around; in particular, they were generated by cosmic rays that came from space and continually struck the earth. They were not actually rays (or radiation); they were mostly particles of various types, including nuclei of various elements and protons and electrons. But when they struck our atmosphere they produced neutrons, and these neutrons could trigger the uranium (or plutonium) prematurely. So the bomb had to be shielded from them.

Nevertheless, the bomb needed a proper and reliable source of neutrons at exactly the right time. A "gun" design was devised to bring together the subcritical pieces and the neutrons all at the same time. But this design was plagued by problems, so another solution was put forward that made use of an implosion. The idea in this case was to construct a sphere of separated plugs of uranium that could be forced together using conventional explosives that would be placed behind them. When all the pieces came together it would explode. Again, there were problems, and just as troubling was the fact that so far a simple reactor had not even been built.

THE FIRST REACTOR

Construction of the first nuclear reactor began in October 1942. A nuclear reactor is a slowed-down version of an atomic bomb; it was needed to verify that a nuclear chain reaction would, indeed, occur, and that a bomb was possible. Enrico Fermi, as the foremost expert in the world on neutrons and neutron bombardment, was put in charge of the project. The reactor was built in the racquet courts beneath the bleachers of the football field at the University of Chicago. It consisted of seventy-six layers of graphite bricks, each of which measured four inches by four inches by twelve inches. Because of this layering, it was referred to as a pile. Scaffolding was eventually constructed around it as it began to grow, so that the upper layers could easily be reached. It was made up by piling two layers of pure graphite bricks, then two layers of bricks loaded with uranium. Cadmium rods were also inserted in the pile. Cadmium is a strong absorber of neutrons, and the cadmium rods would help keep any reaction that occurred under control. They could easily be raised and lowered.[17]

Fermi had two main assistant scientists: Herbert Anderson and Walter Zinn. Each man led a group that worked twelve-hour hour shifts. So work went on around the clock. As the pile was built up, the number of neutrons emitted was carefully monitored. Neutron counters were placed within the pile to do this. A factor called k gave a measure of the number of neutrons that were being generated within the reactor. When k was 1.0, the pile became critical so that the fission reaction was self-sustaining. Fermi wanted to increase it just above 1.0, but he didn't want it to go any larger. If it did, everything could get out of control and an explosion could occur.

Late in the day on December 1, 1942, k was very close to 1.0, and it appeared that criticality would be reached the next day. The next morning a large crowd formed on the balcony that overlooked the reactor. Fermi told his assistant to pull one of the cadmium rods out from the pile slowly. As he pulled it, the clicks in the neutron counters increased rapidly. As he had done throughout its construction, Fermi made some quick calculations using a small slide rule. He then ordered his assistant to pull the rod out a little farther, and again the clicking rate increased.

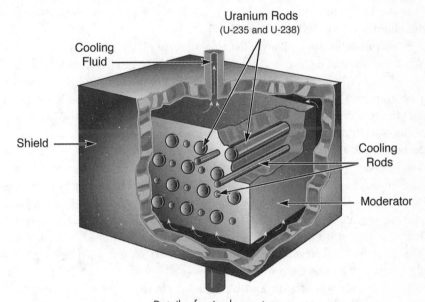

Uranium Rods
(U-235 and U-238)

Cooling
Fluid

Shield

Cooling
Rods

Moderator

Details of a simple reactor.

Everyone was waiting in anticipation, and to their surprise Fermi decided to break for lunch. After lunch they reassembled and Fermi again told his assistant to pull the cadmium rod out of farther. Suddenly the counters went wild. The pile had gone critical. Fermi allowed it to continue clicking wildly for several minutes, then he ordered his assistant to push the rod in to shut it down.

Most scientists now regard this as the beginning of the atomic age. It was the first working nuclear reactor; nevertheless, there was still a long ways to go to get an atomic bomb. But now there was no doubt that it could be built.

THE CONTINUING MANHATTAN PROJECT

Work on the Manhattan Project had started. The major problem was separating U-235 from natural uranium. The uranium nucleus would fission because it was so large and unstable, and it tended to break in half easily. The two isotopes of uranium each had 92 protons, but U-238 had 146 neutrons and U-235, which was the type that fissioned easily, had 143 neutrons. When a neutron was projected at U-235, its nuclei would break down into barium and krypton, and most importantly, when it split, it would release other neutrons that would go on to split other nuclei. The problem was that less than 1 percent of natural uranium was U-235. For the bomb, U-235 was needed, or at least very enriched uranium (uranium that consisted mostly of U-235).[18]

Three methods were known for separating, or enriching, uranium: gaseous diffusion, thermal diffusion, and what was called the electromagnetic method. In the case of gaseous diffusion, natural uranium is passed through some type of porous medium. The heavier nuclei of U-238 will gradually be left behind, and the resulting material gradually increases its percentage of U-235. In this method, uranium is combined with fluorine to form a fluoride gas. Diffusion technology at the time allowed the separation of only micrograms of enriched uranium. So it was obvious that it would have to be done on a very large scale to get enough enriched uranium for a bomb in a reasonable amount of time. The plant was set up at Oak Ridge, Tennessee, in 1943; it was called K-25, and no one working there knew what it was for. Everything was kept secret. Chrysler built the huge diffusers needed, and a problem soon developed. The diffusers had to be built of nickel, and nickel was in short supply, but Chrysler soon devised a way around the problem.

The overall plant was huge, covering an area of two million square feet (half a mile long by four hundred feet wide). The gas passed through ten thousand miles of tubing before it was enriched enough to use in the bomb. About fourteen pounds of enriched uranium were produced from each ton of uranium ore.

The second method of enrichment was called the electromagnetic method. It was discovered at the University of California at Berkeley by Lawrence and his team, and it required the new cyclotron, or atom smasher, that Lawrence had just built. Groves had little confidence in the method because it produced only micrograms of enriched material. Nevertheless, he gave the go-ahead for the work as a backup to the gaseous-diffusion plant, just in case gaseous diffusion didn't work. The electromagnetic plant was also set up at Oak Ridge, and it was called Y-12. Again, it was a huge plant, almost as large as the gaseous-diffusion plant, and, again, none of the workers knew what it was for.

But even with these two programs, things were still going too slow for Groves. He decided to set up a thermal-diffusion plant at Oak Ridge also. Amazingly, it was built in only sixty-nine days. Again, it did not produce very much enriched uranium, but it was soon discovered that if the material from the electromagnetic plant was fed into the thermal plant, the process was much more efficient.

While all this was going on, Groves had another backup. In late 1942 Fermi had shown that a nuclear reactor could be built. And it was soon well known that plutonium could be produced in a reactor from U-238, and that plutonium was also a fissionable nucleus. Furthermore, relatively pure plutonium could be produced at a greater rate than U-235 production. So Groves ordered the construction of three nuclear reactors to in Hanford, Washington. They had the code name X-10. The problem at this stage was that only one relatively small reactor had been built, and the ones in Hanford would have to be huge in comparison, so the technology had to be developed fast. The reactors were built under the direction of Gilbert Church, and strangely, he had no idea what they were going to be used for. He brought in forty-five thousand workers from across the country, and none of them were told what the devices were, or what they were for.

Finally, in early 1945 things began to look up. Considerable amounts of enriched uranium were being produced as well as significant amounts of plutonium. Within months there was enough uranium for a bomb and enough plutonium for several bombs.

While all this was going on, work at Los Alamos was continuing. One by one the problems were being overcome. It was now known how much uranium or plutonium would be needed for a critical mass. Considerable work had been done on both designs for bringing the sub-critical masses together: the gun design and the implosion method. It was, in fact, shown that the gun design would not work with plutonium. Even when U-235 was used for the gun design, it did not appear to work as well as the implosion method. Calculations showed that the implosion would squeeze the masses to super-critical density without the need for a super-critical mass. Furthermore, conventional explosions could be used for bringing the sub-critical plugs together.

Two bombs were developed, referred to as Fat Man (FM) and Little Boy (LB). Little Boy was made with enriched uranium, and Fat Man was made using plutonium. More plutonium than uranium was available at this time, so the initial tests were done using a plutonium bomb.

TRINITY

Things took a strange turn in April 1945. On April 12, Franklin Roosevelt, who had been a strong supporter of the bomb, died, and Harry Truman took over as the thirty-third president of the United States. Strangely, Roosevelt had told him very little about the construction of the atomic bomb, but he did know about the existence of the Manhattan Project. No one knew what to expect of Truman. But, as it turned out, he was up to the job. The war in Germany was over within a few weeks, so the bomb would obviously not be needed there. Japan, however, was a holdout, and it appeared as if it might hold out for a long time.

Before a decision could be made about whether to use the bomb, however, it had to be tested to make sure it actually worked. The test site was called Trinity; it was about sixty miles northwest of Alamogordo, New Mexico, on a desolate stretch of desert. At point zero (the actual bomb site) a 110-foot tower was constructed; the bomb would be placed at the top of the tower. A concrete command center was built approximately ten thousand yards away; several other bunkers were also constructed in the area. A large number of instruments were also scattered around the area to measure the impact of the blast.[19]

The bomb itself was to be a plutonium bomb in which about eleven pounds of plutonium were used. The "ball" of plutonium would be about the size of a small orange. The test was originally scheduled for July 4, but problems developed and it was rescheduled for July 16. Oppenheimer insisted that a dry run (without an explosion) take place before the actual test. It was scheduled for July 14, and to Oppenheimer's dismay a problem was detected. The case for the explosive device was slightly cracked and pitted. Oppenheimer was worried that it would cause a problem, but it seemed too late to call off the test scheduled for July 16. Everything was rechecked; Hans Bethe went through every aspect of the device carefully to make sure that there were no problems. There were, however, a number of uncertainties; the major one was the energy that would be produced by the blast. No one was certain what it would be; estimates ranged from a blast equivalent to forty-five thousand tons of TNT down to one equivalent to only a thousand tons.

The blast was to take place at 5:30 a.m. on the morning of July 16. Within a few hours of time zero a thunderstorm struck and it began to rain. Finally, however, the rain stopped and the sky cleared, so it appeared as if it would go off as scheduled. All observers were equipped with welder's glasses to protect their eyes. The countdown began just before 5:30 a.m. As the countdown reached zero everyone held their breath in anticipation. Suddenly a small bright region erupted close to the horizon. Within a few seconds it had grown into an awesome spectacle: a huge red sphere that was too bright to look at directly. Everyone was silent; then came the blast, followed by a long rumble. At first there was complete silence among the spectators, then several sighs of relief. It had worked. Fermi was quietly busy performing a simple experiment: he dropped several small pieces of paper to see how far they were carried by the shockwave. This would give an estimate of the energy produced. He soon showed that it was equivalent to about ten thousand tons of TNT. News of the success was sent immediately to President Truman.

THE GERMAN BOMB

There was now no doubt: the Americans, with the help of the British, had beaten the Germans to the atomic bomb. But what had happened to the German project? There's no doubt that Hitler wanted super weapons, including the atomic bomb. He boasted about them frequently, but as Germany began to lose the war, he wanted everything as fast as possible, and the V-2 rockets looked like they could be produced much faster than the atomic bomb, so most of his attention was directed toward rocket development. He eventually began to lose interest in funding the atomic-bomb project, so little money was made available. Nevertheless, an active program continued until near the end of the war. By 1943, however, Allied raids on Berlin were increasing rapidly, forcing relocation of the project's major parts to southwestern Germany.

The Americans and British were still worried, however, about how far along the German program was. After all, the Germans had had a better start, with the discovery of fission having taken place in Berlin. Because of this, Grove set up a group of scientists and military officers in September 1943 called the Alsos Mission. Its purpose was to follow the Allies, as they moved through Italy, France, and Germany, to find out as much as possible about the German bomb project and any other similar projects. The group consisted of thirty-three scientists and seven military officers. It was commanded by Colonel Boris Pash with Dr. Samuel Goudsmit as head of the scientific group. They were to capture critical Germans physicists and find any uranium that the Germans might have stockpiled.

For the most part, they followed as closely as possible behind the front lines, but in a number of cases they actually crossed it and came under fire. They soon discovered that about one thousand tons of uranium ore had been shipped to Germany and distributed to several labs in Germany and occupied France. They also found documents and other information at Strasbourg University indicating that there were laboratories related to nuclear research at Haigerloch, Hechingen, and Tailfingen in southwest Germany. There was a problem, however; the Russians were now pushing into Germany from the east, and the French now also had an army that was pushing in the direction of southwest Germany. Groves and other top military brass didn't want the nuclear research sites falling into Soviet hands, or even French hands, before they got to it.

Pash appealed to the top American general to push toward the southwest, but he was told that a deal had already been made with the French. The French would occupy that region, and he would have to get permission from the French to enter it. Pash was annoyed; nevertheless, he set off toward Hechingen and managed to bluff his way past some of the French guards, but his group was stopped before they got there. Again, he had to argue with another French officer, but he was finally allowed to pass.

On the morning of April 24, Colonel Pash and his group finally reached Hechingen. He was surprised to discover that there was still a group of German soldiers in the area, and an hour-long firefight ensued. Finally, his group entered the small town and began looking for the nuclear lab. They soon found Heisenberg's office and lab, and they captured several important scientists, but Heisenberg had already left. His reactor, however, was discovered a few miles away in a cave in the nearby town of Haigerloch. It was beneath a church. The reactor was cylindrical and made up of graphite blocks; the uranium, however, was missing, along with the heavy water that had been used. Nearby, however, the group discovered three drums of heavy water and one and a half tons of uranium ingots buried in a nearby field.

But Heisenberg was still missing. Pushing on, Pash and his men found Heisenberg at his home, waiting for them. The team back at the cave took everything they wanted out of it and set charges to blow it up. Church officials, however, pleaded with them not to detonate the explosives, explaining that the explosion would destroy the church and castle above the cave. So they left it intact.

It was soon obvious that the Germans had made little progress toward the bomb. Heisenberg was still trying to get a nuclear reactor to work, and without it there could be no bomb.

DECISION TO USE THE BOMB ON JAPAN

The Trinity test had shown that the bomb worked. But the war with Germany was over, so it could not be used there. The war with Japan, however, was far from over, although there was no doubt that US forces were winning and that Japan would eventually be occupied. So the question was, should the United States use it, and if so, what cities would should be targeted? As expected, there were arguments from both sides. The Japanese bombing of Pearl Harbor, and the stubbornness of the Japanese at Okinawa and Iwo Jima and other places in the Pacific, showed that *surrender* was a foreign word to them; they would fight to the last man. Furthermore, Tokyo had been firebombed almost to oblivion, yet the Japanese continued to fight. The only alternative, it seemed, was an invasion of the homeland, and few wanted that because it was obvious that a lot of American lives would be lost in the effort.[20]

Many people, however, worried about the ramifications of dropping an atomic bomb. Szilard was one of the most vocal. He tried desperately to meet with President Truman; he even sent a petition to him that was signed by fifty-three scientists. He urged the president to demonstrate the bomb to the Japanese first. Truman apparently looked closely at both sides of the argument and decided to go ahead with the bombing. After all, air raids on Japan using conventional bombs had already produced devastating effects equivalent to twenty thousand tons of TNT. This was about equivalent to the force of a single atomic bomb. And the Japanese still had not surrendered.

Two atomic bombs were therefore dropped, the first on Hiroshima on August 6, 1945, and the second on Nagasaki on August 9. A few days later Japan finally surrendered.

18

THE HYDROGEN BOMB, INTERCONTINENTAL MISSILES, LASERS, AND THE FUTURE

After the development of the atomic bomb, the nature of war changed dramatically. First, an even more powerful bomb, now called the hydrogen bomb was developed. It was, in fact, thousands of times more powerful. Second, with the development of intercontinental missiles, a delivery system was available so that hydrogen bombs could be flown hundreds of miles to a target with the simple press of a button. Finally, with the development of advanced electronics, lasers, satellites, and so on, war began to depend more and more on physics and science in general.

DEVELOPMENT OF THE HYDROGEN BOMB

As we saw in the last chapter, the atomic bomb became possible because it was discovered that heavy nuclei such as that of uranium were unstable and could be easily broken down into two lighter, more stable nuclei. Furthermore, the masses of the two lighter nuclei did not add up to the mass of the heavier uranium nucleus. Some mass had been lost, and it was soon shown that the lost mass was converted into energy. The process in the case of uranium and plutonium was called fission. But there is another, similar process that results in the conversion of mass to energy. It is the process that runs our universe; it allows stars, including our sun, to give off energy, and in the case of our sun, it is responsible for all life on earth. It is called nuclear fusion. In fusion, energy is given off when nuclei come together, or fuse.[1]

Nuclear fusion doesn't occur in heavy elements, however; it only occurs in the lightest elements. In our sun, for example, four hydrogen atoms (actually, just their nuclei) come together and fuse to form helium, and in the process they give up a tremendous amount of energy. The exact details of how this takes place were worked out by Hans Bethe between 1935 and 1938. And it didn't take long after he explained the process of energy generation in the sun for scientists to speculate that a bomb based on the same principle might be possible.

As it turned out, though, it was immediately obvious that the process that took place in the sun would not work for a bomb. It was extremely slow, and the only reason that it worked in the sun was that there was so much hydrogen available. But there are many other fusion reactions that also occur in nature. To understand them, we have to begin with the isotopes of hydrogen; earlier I mentioned that the simplest form of hydrogen has a proton in the nucleus with a single electron whirling around it. It is possible, however, to have neutrons combined with this proton. This doesn't change the element. It's still hydrogen, but the new form with the additional neutron is an isotope. When one neutron is combined with a proton, the isotope is referred to as deuterium; when a second neutron is added, the isotope is called tritium.

Natural water, as you know, consists of both hydrogen and oxygen. The hydrogen in the water that we usually encounter consists of all three isotopes, but only one atom in five thousand is deuterium, and only one in a billion is tritium. So deuterium is relatively rare, and tritium is extremely rare. Scientists determined that the best reactions for a hydrogen bomb were those involving deuterium (D) and tritium (T); they are much faster than the helium fusion that takes place in the sun. In fact, they occur in less than a millionth of a second. But to use them we have to separate D and T from ordinary water, and this is a difficult process. Nevertheless, it appeared as if a bomb could, indeed, be made using D and T.

One of the first to realize that a fusion bomb was possible was Enrico Fermi. He mentioned the possibility to Edward Teller in the fall of 1941, before the Manhattan Project was even organized. Teller was a Hungarian-born physicist who came to United States in the 1930s. He made a number of important contributions to the hydrogen bomb, later becoming known as the father of the hydrogen bomb.

When the Manhattan Project was organized for the development of the atomic bomb, with Oppenheimer as the director, Teller was selected as one of the scientists to work at Los Alamos. Oppenheimer assigned him to a project that involved many long calculations, but he became so intrigued with the possibility of a hydrogen bomb (even though the atomic bomb had not yet been developed) that he neglected the work he was assigned and passed most of it on to his assistant, Klaus Fuchs. (It was later discovered that Fuchs was a spy for the Soviets.)

Teller kept pushing Oppenheimer to start a separate project for the development of the hydrogen bomb, but Oppenheimer refused, angering Teller. Finally, however, Oppenheimer relented and gave Teller permission to look into the possibility. Teller worked on it until the end of the war, and well past it, but made almost no progress. He was determined, however, that such a bomb would work. Finally, in April 1946, a conference was convened in New Mexico to look into the feasibility of a hydrogen bomb. There was increased interest now because the Soviets were known to be working on their own atomic bomb, and it was possible that they were also considering the construction of a hydrogen bomb.

In August 1946, President Truman signed a bill that established the Atomic Energy Commission, which was to look into the use of atomic science and technology, not only for weapons, but also for peacetime use. Within a couple of years it became known that Klaus Fuchs had passed many of the secrets of the hydrogen bomb to the Soviets, and it became clear that they would likely soon be developing a hydrogen bomb. Many military people began to worry, and in January 1950, President Truman announced that it was now important to go ahead with the development of a hydrogen bomb. There was, however, a strong difference in opinion among the scientists who were likely to be involved in the project. As expected, Teller was jubilant, and others, such as Ernest Lawrence, were also strongly for it. But Oppenheimer advised caution; he was worried about the consequences of such a weapon, as were Bethe and several others.

Nevertheless, a "crash program" to develop what was called the "super" at that time, went ahead. Many of the scientists who had earlier been involved in the Manhattan Project were called back to Los Alamos.

THE ULAM-TELLER BREAKTHROUGH

By this time Teller had already spent several years trying to devise a model that would work, but he hadn't come up with anything that could be taken seriously. It almost seemed as if such a weapon was not possible. One of the new people now working on the project was a Polish mathematician, Stanislaw Ulam, who had come to the United States in 1935. He had worked at the Institute for Advanced Study at Princeton for a while, and then in 1943 he joined the Manhattan Project, where he worked with John von Neumann. And in 1946 he went to Los Alamos to work on the development of the hydrogen bomb.

His job was to look into the feasibility of using either a D-D reaction or a D-T reaction to trigger the fusion reaction needed for the bomb, and to come up with an appropriate design. Various designs had been tried, but nothing seemed to work. At this point it was known both that a tremendous amount of heat (twenty to thirty million degrees) was needed to trigger a fusion reaction and that an atomic bomb could be used to create such heat. But everything Ulam tried appeared to have problems. In December 1950, however, he stumbled on an idea that he was sure would work. Basically what was needed was a way to increase the compression of the hydrogen in the bomb by several magnitudes. An atomic explosion could be used to create an implosion that would compress the hydrogen, but a simple implosion didn't appear to be enough. Ulam decided that several explosions were needed. In essence, one bomb would be used to set off a second bomb, and the second bomb would set off a third. This was referred to as staging. He was sure the idea would work, but he kept it to himself for many months while he developed and perfected it.[2]

He finally decided to tell Teller about it, even though he didn't have a good relationship with Teller and was worried about Teller's reaction. Teller was not immediately convinced that it would work, but as he continued to study the idea he realized that it was an important step forward. Ulam suggested that the hydrodynamic shock, or possibly the neutrons from the fission explosion, could be used to create an implosion that would compress the hydrogen sufficiently. After studying the possibility for a time, Teller realized that the x-ray radiation would reach the hydrogen before the shockwave or the neutrons, and that it could be used to create the implosion that would be needed to trigger the thermonuclear explosion. And indeed it appeared to be the best solution. Teller and Ulam submitted a joint paper on what soon became known as the Ulam-Teller Design. For several years, however, Teller tried to play down Ulam's contribution, and there was considerable friction between the two men.[3]

THE FIRST TEST: MIKE

The next step was to build a bomb based on the Ulam-Teller Design to see if it would work. And indeed, work began on the project relatively soon. In reality this first bomb was not a bomb, as we know it; it was far too large to carry in an airplane. The basic parts of the device would be manufactured in United States and taken to a remote location in the Pacific Ocean, about three thousand miles west of Hawaii. The test, codenamed Ivy Mike, was conducted at Enewetak Atoll, a ring of forty small islands about forty miles long and ten miles across.[4]

A committee called the Panda Committee had been set up to look into the development and testing of the bomb. Its members were given a year to design and deliver the bomb, even though there were still many problems to overcome. One of the major problems was deciding which fusion reaction to use: D-D or D-T. It was finally decided that the D-D reaction would be both easier and more economical. But there was a problem in relation to how the deuterium would be stored. Deuterium has a boiling point of 417 degrees below zero Fahrenheit, so it had to be kept in a liquid state at extremely low temperatures. This required that it be stored in a cryogenic system—a large Dewar (or vacuum flask) that would keep it at a very low temperature. In addition, the device would require a fission bomb to trigger the hydrogen fusion, and at this time fission bombs were still relatively large. The radiation from this explosion would then be channeled into a secondary that contained liquid deuterium. The overall bomb was in the form of a cylinder, with a stick of plutonium at its center that would act as a "spark plug" for initiating the fusion reaction.

Formal assembly of the system, called "Mike," began in September 1952. The bomb itself was placed at one point along the atoll, and several monitoring stations were set up at other points for measuring the energy output of the blast. In addition, a large number of ships were stationed around the atoll, and a number of aircraft were in the air, loaded with measuring equipment. In all, there were over four hundred scientific stations with measuring instruments of various types around the blast site.

By September 25, everything was ready; zero hour was to be 7:15 a.m., November 1. The "firing room" was actually about ten miles away, aboard a ship called the *Estes*. The power of the blast amazed almost everyone; again, as in the case of Trinity, no one was certain how powerful it would be. As it turned out, it was considerably more powerful than anticipated. Almost immediately a blinding, white-hot fireball formed on the horizon. It was three miles across, compared with the fireball of the Hiroshima blast, which was only a tenth of a mile across. Within two and a half minutes the cloud caused by the shock wave had reached an altitude of one hundred thousand feet, and it continued to billow out, eventually forming a huge canopy thirty miles across. The blast literally vaporized the entire island on which Mike had been staged, leaving a crater two hundred feet deep and more than a mile across. The energy of the blast was determined to be equivalent to 10.4 megatons of TNT. This was by far the largest man-made explosion ever to occur on earth.

PHYSICS OF THE HYDROGEN BOMB

Let's look now at how and why the hydrogen bomb works. In many ways it is much more complex than the atomic bomb. But without an atomic bomb it wouldn't work, so the atomic bomb had to come first. As we saw in the above section, it is, in effect, a staged radiation implosion that provides the required temperature (about 50,000,000°) for fusion reactions to occur.

For fusion reactions we need deuterium or tritium, and, as we saw earlier, they are relatively rare and must be separated from natural water. Reactions using both deuterium (D) and tritium (T) can be used, but tritium is much more expensive to produce, so scientists tried to avoid using it directly. But even though deuterium is much more plentiful, it is difficult to store and must be in a liquid state at very low temperatures, as was done in the case of Mike. This was eventually overcome by combining deuterium with lithium to produce lithium deuteride, which is a stable solid that is much easier to handle than deuteride. All modern hydrogen bombs now use lithium deuteride.

Basically, what is needed is an implosion of tremendous power that is able to compress the fusion fuel to densities high enough so that the fusion reaction can occur. The required density is at least a thousand times the fuel's normal density.

The simplest hydrogen bomb is a two-stage device, and this is the only type we will discuss here. It is possible to go to three stages, but most bombs use the two-stage configuration. The Soviets built a three-stage device, but little is known about it. As we saw earlier, the Ulam-Teller Design utilizes x-rays for compression because they move particularly fast (the speed of light) after the primary (the atomic bomb) is ignited. Shockwaves and neutrons are also emitted in the explosion, but they are too slow to use.

One of the critical aspects of the hydrogen bomb is precise sequencing of the stages. If anything is out of sequence, the bomb will not work. So timing is crucial. The overall bomb is in the shape of a cylinder (later bombs had a more elliptical shape). The primary (trigger) is at one end, and the secondary (the fusion device) is at the other end. The secondary generally takes up more space than the primary. The secondary is also in the form of a cylinder; it is smaller, however, than the outer cylinder, so there is a space between the two cylinders. This space is called the radiation channel. The fusion fuel, which is lithium deuteride, takes up most of the space in the secondary. The outer edge of the secondary is made up of U-238, and it is referred to as the pusher/tamper. When triggered, it pushes inward on the secondary fuel. At the center of the secondary, running down the axis, is a rod about one inch across that is made either from plutonium-239 or U-235. It is referred to as the spark plug.[5]

The area between the secondary and the outer cylinder is filled with plastic foam. And there is a large curved shield in the front of the secondary to prevent the fusion material from being triggered prematurely. When the primary (an atomic bomb) explodes, x-rays from it fill the radiation channel. This area is filled with plastic foam that becomes ionized after the initial explosion; it helps to control the explosion. It is important that the tamper/pusher on the outer section of the secondary is not heated unevenly or too fast. An "equilibrium" condition is needed so that the energy throughout the region is uniform.

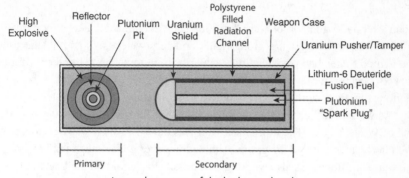

Internal structure of the hydrogen bomb.

As the explosion proceeds, the outer layer of uranium on the secondary finally fissions and an implosion occurs. The implosion compresses the fusion material, producing neutrons in the process. These neutrons trigger the uranium (or plutonium) rod at the center of the secondary, and it explodes. As a result, the fusion fuel is compressed from the top and the bottom. It therefore quickly reaches a temperature high enough for fusion reactions to occur. At this point the fuel has a density more than one thousand times its original density. Some tritium is generated in the fusion reaction, so in practice both D-D and D-T reactions occur.

It's easy to see from this that the hydrogen bomb derives its energy, or explosive power, from both fission and fusion. So the overall blast can be considered to be both a fission and a fusion blast, which may seem to be unimportant, but there is a significant difference in the two blasts. The radioactivity that spreads out after a bomb is exploded comes from the fission blast, whereas the fusion blast is "clean" in this respect. So when someone talks about producing a clean, or radiation free, bomb, they are referring to one that only has a small fission blast. It is, in fact, possible to build a relatively clean nuclear bomb.

The largest American hydrogen bomb had an explosive power of about fifty megatons of TNT. The Soviets exploded one that was even more powerful than this. Bombs can, in fact, be made more powerful by adding additional stages to them. And, as discussed previously, three-stage bombs are believed to have been built by the Soviets. What is particularly important in relation to the power of a hydrogen bomb is that, in theory, there is no limit to how powerful it can be built. There *is* a limit in the case of an atomic bomb.

LONG-RANGE MISSILES

Soon after hydrogen bombs were developed, it was realized that a better delivery system was needed. At first, long-range bombers were used, and at that time the United States had overwhelming superiority in long-range bombers. But as rockets became more sophisticated and their range was expanded, it became obvious that they would be much more appropriate as a delivery system.

As we saw earlier, the Germans developed the first ballistic missiles near the end of World War II. The most successful was the V-2, which was developed by Wernher von Braun and his group. Although it never got as much publicity after the war, von Braun was also working on a missile that could hit the United States. It was called "Project America." With a far greater range, Hitler had hoped to use it on centers in America. Fortunately, it was never developed and used.

When the war was over von Braun and many of the other German rocket scientists came to United States, but some went to the Soviet Union. And very quickly the Cold War developed, with both nations stockpiling large numbers of nuclear weapons along with large numbers of long-range missiles to deliver them. Several projects were, in fact, initiated. At first they were merely extensions of the German V-2 program, but improvements came quickly, with the Soviets soon gaining an extensive lead. In August 1957, the Soviets launched the first intercontinental ballistic missile, which they called the R-7. And within a short time they also launched the first orbiting satellite, Sputnik—much to the shock of the Americans. This was followed by the launching of the first human into space, cosmonaut Yuri Gagarin.

The United States immediately developed a "crash" program in an attempt to catch up, and when the Russians detonated their first hydrogen bomb in 1953, the urgency increased. Plans for the development of the Atlas rocket were initiated in 1954, but it was not until 1958 that it launched successfully.

Soon there were two programs. One was for the development of intercontinental ballistic missiles (ICBMs) that could be used to carry nuclear weapons.[6] The other, initiated by President Kennedy about the same time, was called the Apollo program, and it used Saturn rockets. The Apollo program had the goal of taking a man to the moon. Many of the earlier rockets, such as the Atlas, Redstone, and Titan, formed the basis of both this program and the ICBM program.

An ICBM is a ballistic missile with a range of more than three thousand five hundred miles, and ICBMs were usually designed to carry nuclear warheads. Many now have a range of up to twelve thousand miles. Modern ICBMs usually carry multiple independent re-entry vehicles, or MIRVs, each of which carries a separate nuclear warhead. This makes a single ICBM much more effective and deadly because it can hit several targets at once. MIRVs have become possible because nuclear warheads (hydrogen bombs) have become much smaller over the years. In addition, the rockets themselves have become smaller and now have a much greater range.

All early ICBMs were launched from very vulnerable, fixed, above-ground sites that could be easily attacked. This changed significantly over the course of the Cold War. Many were put in protected silos, mostly in northern states. In addition, they were now small enough that they could be launched from heavy trucks and railroad cars, which made them quite maneuverable. The most effective launch sites, however, are nuclear submarines. Once nuclear reactors were developed and perfected, they were soon used in submarines, and they proved to be particularly effective in them. Where early diesel-electric submarines had to surface frequently, nuclear submarines could stay submerged for months on end. And there was almost no need to refuel them; enough fuel for a sub's reactor for up to thirty years could easily be carried aboard the submarine. In some cases the reactor generates electricity that is used to power the propeller, and in other cases a reactor creates steam that drives turbines. Nuclear submarines are, however, very expensive to build, and because of this, only a few nations have them.

All American nuclear submarines are now equipped with ballistic missiles that have an intercontinental range. The biggest advantage of the submarine in this respect, however, is that it is highly maneuverable, relatively difficult to detect (although subs can be detected with sonar), and big enough to carry several MIRVs.

Soon after ICBMs with warheads were developed, several nations began to consider how they could be countered. In particular, could a missile be developed that had the ability to shoot down an incoming ICBM? Such systems were referred to as anti-ballistic missile (ABM) systems. The first study into the possibility of such a system was actually made as early as the later part of World War II by Bell Labs. The British had been bombarded by V-1 rockets, and later by V-2s, and they were looking for a defense. The V-1s were not ballistic, and British fighter planes and land-based artillery were able to shoot some of them down. But when the V-2 ballistic missiles appeared, there appeared to be no defense because of their high velocity and altitude. The Bell Labs study concluded, in fact, that it was not possible to shoot down a V-2 rocket. But that was before the advent of high-speed computers, and by the mid-1950s several nations were, indeed, considering the possibility of ABM systems.

Such systems are now divided into two classes: those that are directed against ICBMs and those that are directed against smaller rockets. At the present there are only two systems that can intercept ICBMs, since they are a much greater challenge than smaller rockets. The United States has developed what is called the ground-based midcourse defense (GMD). It consists of interception missiles along with a radar system to detect the incoming ICBMs. This system has been tested extensively over the years, with a mixture of successes and failures. It is still being worked on. The United States now has several smaller, short-ranged tactical systems that are more effective.

OTHER WEAPONS: LASERS

Another important modern weapon of war is the laser. When the first lasers were developed in the 1960s it was thought that they would soon become serious weapons, perhaps replacing guns. After all, Buck Rogers and many other early science fiction characters used "ray guns," and it was believed that they would soon become reality. As it turned out, though, they have not replaced traditional guns, but recently they have been used to knock down drones and possibly disable small ships or boats. They have also been used extensively for marking targets to determine their range.[7]

Despite their limited use as weapons of war so far, lasers do have considerable potential in that area, and they have been used extensively in everyday devices. They are used in DVD players and laser printers, as barcode scanners in stores, and their use in medicine has created a revolution in surgery. Furthermore, they are being used extensively in industry for cutting and welding.

The origin of the laser can be traced back to an early paper by Einstein. In an even earlier paper, Niels Bohr of Denmark postulated that atoms consisted of nuclei (protons) with electrons whirling around them in various discrete orbits, corresponding to various energies. We can, in fact, draw a simple energy diagram for an atom. Bohr mentioned the possibility of electrons jumping back and forth between these energy levels, but it was Einstein who put the idea on a firm basis.

In the diagram we see several energy levels with electrons in some of the levels. When an electron absorbs a photon of light it moves to an upper, or excited, level; that is, to an orbit more distant from the nucleus. Usually it only

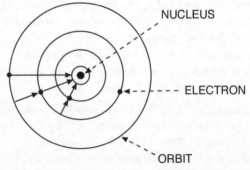

Basic structure of an atom.

stays there for a short time before it jumps back down to the original level (called its ground level). When it moves from an excited state to a lower state it emits a photon of light. This is referred to as spontaneous emission. Einstein also introduced the idea of stimulated emission; in this case the electron is already in an excited state. If a photon is directed at this electron, it can be stimulated to fall to a lower energy state, but it does not absorb the photon. In fact, it emits another photon as it falls down to the lower state so that we have two photons coming out in the process. And of particular importance, they both have the same wavelength, and they are in phase.

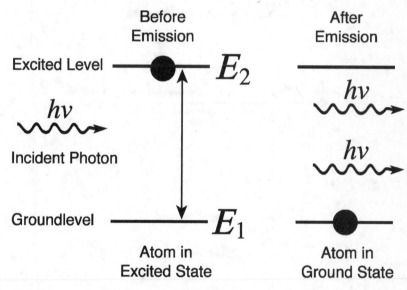

Stimulated emission.

It was an interesting phenomenon, but for several years no one took any interest in it. During World War II, however, radar was developed and used extensively, and there was considerable interest in further developing it after the war. One of the areas of interest was the possibility of a microwave amplifier; in other words, a device that would increase, or amplify, microwaves. Joseph Weber of the University of Maryland became particularly interested in the device. After studying the problem in detail he came to the conclusion that it might be possible to build an amplifier using stimulated emission. He pointed out, however, that what is called a "population inversion" would be needed. Such an inversion occurs when high-energy levels of an atom contain more electrons than lower levels. This is not the normal situation; the electrons in an atom are usually distributed so that most are in lower energy levels, with fewer in upper energy levels.

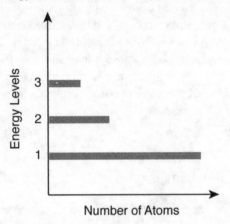

A typical energy diagram showing the number of electrons in each level.

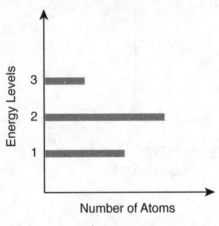

A population inversion.

But how could a population inversion be created? Obviously an energy source would be required to force the electrons to higher energy levels, and appropriate energy sources were soon found. We now refer to them as pumps.

Weber designed a microwave amplifier that he thought might work, but he didn't build it. This was left to Charles Townes of Columbia University. Townes was also studying microwaves and looking into the possibility of an amplifier. He decided to set up a population inversion with what he called a resonant cavity, which is a box with reflecting walls. He devised a method for pumping the electrons within this resonant cavity up to excited states, and by doing so he succeeded in creating a population inversion. In addition, he devised a method for allowing the electrons to suddenly fall to the ground state. The radiation they gave off when this occurred was "coherent" microwave radiation; in other words, the wavelengths were all lined up and had the same phase and frequency (see diagram). In the process he produced the first of what is now called a maser (whereas a maser uses microwave radiation, a laser uses visual light).

Charles Townes.

Soon after he created the maser, Townes began to look at the possibility of a similar device that used optical waves or visible light. This did not prove to be easy. Optical photons are quite different from microwave photons, and Townes worked on the device for several years before he managed to build one. The new device, called the laser (short for light amplification by stimulated emission of radiation), has now overshadowed the maser because it has many more uses in modern society.

The basic principle of the laser is similar to that of the maser. A laser creates a beam of light in which all the photons are coherent. In an ordinary light beam the photons are of many different wavelengths (white light is composed of all colors, each of which has a different wavelength), and the waves are not lined up; as a result, they easily knock one another out of the beam so that the beam cannot be sharply focused. In a laser beam the photons (or waves) are coherent and of the same frequency so that they *can* be sharply focused.

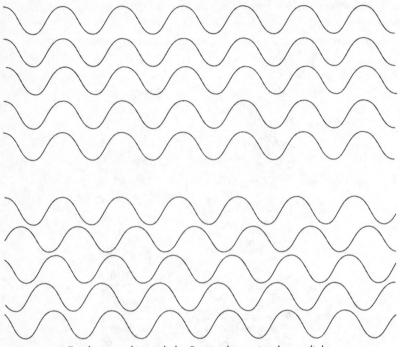

Top beam: coherent light. Bottom beam: incoherent light.

As in the case of microwaves, a resonant cavity is also needed here; in the laser, though, it is usually called an optical cavity. The medium within the optical cavity is called the gain medium; it is a material that has the properties needed for amplification of the light by stimulated emission. For this amplification a pump is needed; it is usually an electrical circuit or a flash lamp. Mirrors are placed at either end of the optical cavity, one of which is partially transparent so that some light can pass through it. Light within the cavity reflects back and forth through the gain medium and is amplified each time it makes a pass. This medium is, in essence, a population of atoms that has been excited by an external source. The medium itself can be liquid, gas, solid, or plasma.

The gain medium is "pumped" to an excited state; in other words, the atoms within it are in the excited state after the pumping occurs. Eventually a population inversion is achieved in which higher energy states are more densely populated than lower energy states. The reflected beam grows in intensity until finally it is powerful enough to break through the partially reflecting mirror. What emerges is a coherent laser beam.

Townes, along with his student Arthur Schawlow, was the first to design a workable laser, but they did not build it. They did, however, publish a paper on it, and they filed a patent for the idea in July 1958. A research worker at TRG Incorporated by the name of Gordon Gould was also working on a similar device. Gould tried to patent his device in April 1959, but his application was turned down, even though Gould had described the construction of his laser in a notebook prior to Townes and Schawlow filing for their patent. Several court cases followed, and it took years to settle them.[8] The two groups are now credited with having invented the laser independently.

The first person to actually build a working laser, however, was Theodore Maiman of Hughes Research Laboratories in California. His device was quite different than that of Townes, Schawlow, and Gould; they had designed a device using gas as the gain medium. Maiman used a ruby rod with a helical flash lamp wound around it that acted as a pump.

The next step was, of course, to use lasers as weapons of war. Laser-like devices such as ray guns had been used in science fiction for years. It turned out, however, to be much more difficult to make laser weapons than expected, and there's little chance that a laser-like weapon will replace small arms in the near future. The main problem is that lasers require a huge power source, and because of this, there are serious engineering problems. Larger weapons, however, are possible, and the navy has recently built one that could disable an enemy ship and knock down enemy drones. The biggest advantage of a laser such as this is that it doesn't require expensive ammunition. However, the laser itself would be relatively expensive.

One form of laser that appears to have considerable potential is the x-ray laser. It produces a coherent beam of x-rays rather than an optical beam; as a result, it has much more energy. It was considered part of the Strategic Defense Initiative that was proposed in 1983 (sometimes referred to as "Star Wars"). Such lasers were to be powered by nuclear explosions. Tests eventually showed, however, that they were not feasible.

TRANSISTORS, MICROCHIPS, AND COMPUTERS

Many scientific breakthroughs have led to important developments in weaponry, but nothing approaches the invention of the transistor. All electronic devices now use transistors in one form or another, and hardly a form of weaponry exists that doesn't use electronics in some way. The electronic age came about early in the twentieth century with the invention of the triode, or vacuum tube. It gave us radar and many other electronic devices. But it was fragile in many ways, and relatively large. When John Bardeen, Walter Brattain, and William Shockley developed the transistor at Bell Labs in late 1947, however, the world of electronics underwent a revolution. Tiny radios, calculators of all types, and powerful computers soon followed. Today most transistors are actually found in integrated circuits, or microchips, as they are frequently called; nevertheless, it was the invention of the transistor that started the revolution.[9]

A transistor is a device that can amplify, or switch, incoming electronic signals. It was developed by physicists working in solid-state physics. As the name implies, solid-state physics deals with solids. And, as you no doubt know, solids come in many varieties. Some are good conductors of electricity; others are insulators (nonconductors), and there is a group in between called semiconductors. Semiconductors have proven particularly important because solids of this type have made transistors possible.

To understand things a little better, let's look at the atomic structure of metals and semiconductors. We'll begin with a gas. The atoms of a gas have a nucleus with a number of electrons whirling around it in various energy levels. Assume that we apply pressure to the gas or lower its temperature. What happens to it? The atoms begin to move closer together and eventually the gas turns into a liquid as the atoms get closer and closer. At this point the energy levels of the various atoms are still completely separated, but as you continue applying pressure (or lowering the temperature), the liquid becomes solid, and the energy levels of the individual atoms begin to overlap. They will create what are called energy bands, which are continuous regions of energy.

The exact way these bands form depends on the particular material being compressed or cooled. If you could look at these energy levels closely you would see that some of them contain electrons, and some are empty. There are also gaps between the bands. In most metals and semiconductors there are, in fact, two major bands with a gap between them. They are referred to as the valence band and the conduction band. The size of the gap between the two bands determines whether they are metals, semiconductors, or insulators.

A current, as we saw earlier, is a group of electrons moving through the lattice created by the atoms of a metal or a semiconductor. In effect, the electrons jump from atom to atom. To move through the lattice, however, they have to have enough energy to overcome the "gap energy." In other words, they somehow have to acquire enough energy to jump from the valence band up to the conduction band. Semiconductors have relatively small gaps, so it doesn't take a lot of energy for electrons in the valence band to jump up to the conduction band. Conductors such as copper, on the other hand, have little or no gap, and electrons flow very easily when a small voltage is applied.

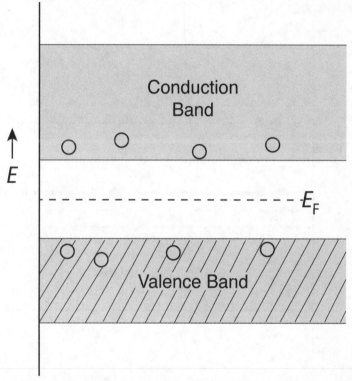

Conduction and valence bands. Note the gap between them. E_F is called the Fermi levels.

Two of the most important semiconductors, as far as electronic systems are concerned, are germanium and silicon. What makes these semiconductors particularly valuable is that they can be "doped" with impurity atoms such as boron and phosphorus. Impurity atoms have either a deficit or excess of valence electrons (valence electrons are responsible for the electrical conductivity of different elements). Doping is the process of inserting these impurities, which make new energy levels available within the gap, either just below the conduction band or just above the valence band. The levels just below the conduction band are created by donor impurity atoms; the levels just above the valence band are created by acceptor impurity atoms. Semiconductors doped with donor impurities are called n-type. Those doped with acceptor impurities are called p-type. When an electron jumps to an acceptor level it leaves a "hole" in the valence band, and this hole acts like a positive electron.[10]

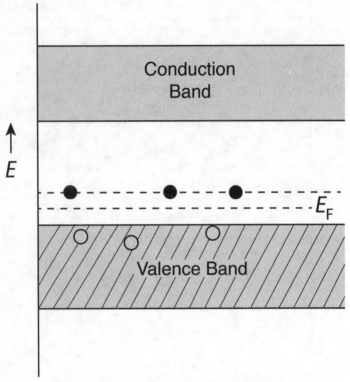

Energy-level diagram of a semiconductor with electrons
in the acceptor levels and holes in the valence band.

Armed with this information, Bardeen and Brattain began looking into how semiconductors could be used in electronics. One of the simplest electronic devices at the time was the rectifier—a device that would allow current to flow only in one direction. Having decided to look into the possibility of creating a rectifier using semiconductors, they found something that was of even more interest: a simple form of amplification. Amplification is an increase in the signal; it can be an increase in current, voltage, or power. In their experiments Bardeen and Brattain achieved current and power amplification but not voltage amplification. Their first device used point contacts on the surface of the semiconductor.

Bardeen and Brattain continued to improve their device, but there were several problems with the contact probes. One of the major ones was that there was a surface layer on the semiconductor that appeared to be causing problems. William Shockley, who was the leader of the group, now became more involved. He suggested that a three-layered semiconductor structure would work just as well and would be simpler. This would, in essence, be two p-n junctions placed back to back to form either a p-n-p or an n-p-n device, which we now call a transistor.

Several connections can be made to a transistor; usually an input signal through two connections is amplified and the resultant, or output signal, is obtained through two other connections. Over the years the size of transistors has decreased significantly; as a result they are now incorporated into very small circuits of various types. They soon became the central device for computers, and with increasing technology they became even smaller and smaller. As a result, computers also became very small.

Eventually most transistors were integrated into tiny circuits called microchips. Tiny wafers began to hold hundreds, then thousands and even hundreds of thousands of tiny transistors and other electronic components. And surprisingly, as microchips became smaller they also became more reliable. Today literally billions of transistors can be placed on a tiny microchip. As a result, computers of all types now surround us in an incredible variety of devices, and they have revolutionized the weapons of war. They are found in tanks, airplanes, guided missiles, rockets, many types of guns, and almost all types of bombs.

SATELLITES AND DRONES

We don't normally think of satellites as weapons of war, and so far they haven't been involved in direct fighting, although in theory they could be equipped with many different types of weapons, including lasers of various types, particle-beam weapons, and even missiles. As we saw earlier, Sputnik was launched by the Soviets in 1957. Explorer 1, the first American satellite, was launched the following year, but for several years the United States trailed the Soviet Union in space technology. Although satellites are used for many commercial purposes, including transcontinental TV broadcasting, long-distance telephone transmission, weather prediction, and GPS navigation, one of their major uses is for spying, and we will direct our attention primarily to spy satellites.

Within a few years after Sputnik, both the United States and the Soviet Union were launching satellites for spying. Early spy satellites recorded data then ejected it in canisters that had to be retrieved. It wasn't long, however, before radio came into use as a means of retrieving the information. The first series of spy satellites launched by United States in 1959 was called Corona. Since then a large number of spy missions have been initiated, as spying techniques have become more and more sophisticated. Many other nations, including Israel, the United Kingdom, France, Germany, and India, are now launching their own spy satellites.

This sky is now full of spy satellites, most orbiting overhead at altitudes of one hundred to two hundred miles. They travel at approximately 17,500 miles per hour, taking snapshots of millions of different items of interest to the military and the Central Intelligence Agency. They are, in effect, giant digital cameras pointed at the earth. Everyone has heard of the amazing discoveries made by the Hubble Space Telescope, with its giant mirror. As it turns out, the United States has telescopes in satellites that are now just as large and powerful as Hubble, but they're pointed toward the earth. They are referred to as Keyhole-class (KH) spy satellites, and they provide very high-resolution images; they can, in fact, resolve objects down to five or six inches.[11]

But high resolution isn't their only feature. The newer satellites can now take pictures in stereo (side-by-side images at a slightly different angle) that, with the help of computers, can give three-dimensional images. In addition to this, radio images and infrared images can be obtained. Infrared imagery is particularly helpful because it allows the cameras to see through clouds and also at night. Furthermore, intelligence collecting has now become so refined and fast that a single radio or cell phone can be located and pinpointed almost immediately, and orders for targeting its owner can be issued within minutes.

Most satellites, in fact, now contain extensive, fast computers that can crunch huge amounts of data within a fraction of a second. This data can be quickly transmitted down to an operation center on earth.

In addition to satellites, unmanned aircraft are now also being used extensively. They are usually referred to as drones. Drones have been used extensively in the wars in Iraq and Afghanistan. The two major types now being used by United States are the MQ-1 Predator and the MQ-9 Reaper (but others are also in use). They are referred to as UAVs (unmanned aerial vehicles) or RPVs (remotely piloted vehicles). And there's no doubt that they are changing the nature of modern aerial combat and combat in general. Their main advantage, of course, is that a pilot is no longer in danger. Nevertheless, the craft can still inflict considerable damage to an enemy. Another important advantage is that they are much cheaper to build than conventional fighter planes. The predator is only about twenty-seven feet long.[12]

The "pilot" of these drones is usually thousands of miles away. For the ones being used in Iraq, Afghanistan, and Pakistan, the pilot is usually located at a military installation in the United States, where he or she is positioned in front of a screen that shows what a pilot in a plane would normally see, and manipulates the drone as if sitting in the drone's "cockpit." Furthermore, the pilot is able to communicate with troops on the ground below the drone. In particular, he or she can give them information about the position and capability of the enemy.

A predator drone.

Most drones are considerably smaller than fighter planes, and they are not as well equipped. Predator drones usually have no armaments, since they are used mainly for spying; Reapers, however, are equipped with missiles. The British are designing a model they call Taranis, however, which will be about the size of a fighter plane. It will be equipped with weapons of several types, and it will be capable of defending itself from attacks by other aircraft. The Israeli air force also has drones called Hermes 450s, which are equipped with missiles. Many of the countries with drones use the Sperwer, which is produced in France. It is capable of twelve hours of sustained flight, and it is equipped with various electrical-optical devices including infrared and radar sensors; it also carries missiles and antitank weapons.

FUTURISTIC WEAPONS OF WAR

You merely have to look at science fiction to see all kinds of futuristic weapons. But how many of them are practical or even possible? Some of them will no doubt eventually be used in war, but most will not. Let's look at some of the futuristic weapons that may one day make it off the drawing board.

One of the most interesting is called the e-bomb, and although it could be devastating to a civilization, it does not kill. The idea for such a bomb first came in 1960 when the first hydrogen bombs were set off. One of the phenomena measured was the intensity of the electromagnetic pulse generated by the blast. Scientists soon determined that this pulse was felt at a considerable distance — as far as nine hundred miles from the blast. Furthermore, the blast compromised the functioning of instrumentation in airplanes miles away.[13]

Scientists didn't give much thought at first to the danger associated with the electromagnetic pulse that was generated. But they wondered *how* it was generated. And they soon found out. In a nuclear blast, large numbers of gamma rays are generated, and they, in turn, produce high-speed electrons, some of which become trapped in the earth's magnetic field. These electrons produce powerful electric and magnetic fields, which in turn produce extremely high currents and voltages in any type of electronic or electrical equipment. In effect, all electronic equipment is destroyed by these pulses, including all computers, communication equipment, and telephones, as well as electrical systems in cars, airplanes, and so on. Such a blast could bring society to a standstill and cause billions of dollars in damage.[14]

The military has been looking into how to generate brief but powerful electromagnetic pulses. It would be inconvenient to have to set off a nuclear bomb to create them, and indeed they can be produced relatively easily without a bomb. An explosive packed into the interior of a large copper coil is all that is needed. The instant before the explosion, the coil has to be energized by a bank of capacitors to create a magnetic field. The explosion creates a moving short circuit, which in turn compresses the magnetic field. The result is an intense electric pulse that immediately propagates outward from the device.

The pulse from such a device could paralyze a large fraction of the United States. If detonated about 250 miles above a central state such as Kansas, it could knock out all the electronic equipment and electronic devices throughout most of the United States.

Is there any defense against such a system? Indeed, it is possible to shield electronic systems, but such shields are costly, and they are not likely to be practical for many years.

What about other systems? We talked earlier about x-ray lasers, and considerable work has gone into trying to develop them. There is, however, a serious problem with the concept, as compared to an optical or microwave laser. The lifetime of the excited electrons it produces is very short, and there are difficulties in building x-ray mirrors. Because of this, x-ray laser beams generally have poor coherence, and it is difficult to get around this. The best alternative appears to be the use of highly ionized plasma as the active media. This idea shows some promise, but effective weapons using x-ray lasers have so far not been built.

Another similar approach is to use high-energy beams of atoms or subatomic particles as weapons. Such beams are of course being generated every day all around the world. They are the beams created by accelerators of various types, including cyclotrons and linear accelerators. The technology associated with particle accelerators is, of course, well known. The particles that are usually accelerated are electrons, neutrons, positrons, protons, and ionized atoms. Charged particles are difficult to focus and keep in a narrow beam because they repel one another, so the best particle for a weapon would be neutrons. Beams of this type have several advantages. First, the particles travel at speeds close to that of light. Second, they can be produced with very high energies. Work on such beams is currently going on in several places, including the Ion Beam laboratory at Kirtland Air Force Base.

Returning to less exotic weapons, researchers are also developing very sophisticated grenade launchers. The XM25 grenade launcher is equipped with a laser rangefinder and an on-board computer. It is designed so that the grenade can be guided to the target using a laser beam. The grenade will then explode in midair directly above the target.[15] "Smart bullets" may also be on the way. These are bullets that are maneuverable in flight and are controlled by a guidance system. They will be particularly effective if the target is moving.[16]

Robots of various types have also been considered for years as possible weapons. They have already been shown to be useful for clearing mines. However, the Modular Advanced Armed Robotic System (MAARS®), which is being marketed by QinetiQ North America, is an unmanned ground vehicle that is operated remotely. It has onboard cameras, motion detectors, a microphone, and several other devices. It is propelled using continuous tracks, like those used on tanks, and it can be deployed for reconnaissance, surveillance, and target-acquisition purposes. It can also carry a weapon system.[17]

Finally, another possible weapon is worth mentioning, even though it may seem farfetched. Earlier we saw that drones and satellites are being used for spying and other purposes. As strange as it may seem, sensors are being developed that can decipher the electromagnetic waves in the human brain. A satellite or drone equipped with such a device might one day be able to "read the mind" of an enemy on the battlefield.

Some of these possible weapons of the future may seem like products of pure fantasy, but imagine the wonder a warrior of the Persian Empire might have experienced if he'd been told about the muskets used during the Thirty Years' War. What might one of Napoleon's soldiers have thought if told about the armed airplanes and submarines of World War I? To the citizens of Nagasaki and Hiroshima atomic bombs would have no doubt seemed fantastical, that is until August 1945, when atomic explosions suddenly became all too real. By increasing our understanding of the physical laws of nature, humans have learned to produce weapons of ever-greater destructive power. As physicists further expand our knowledge, it is almost certain that our weapons of war will continue to progress. The great hope for the twenty-first century and beyond is that rather than increasing the carnage of war, such progress will instead promote the development of precise, nonlethal weapons that ultimately enable the resolution of conflict without the staggering human slaughter that became too common in the twentieth century.

NOTES

CHAPTER 1. INTRODUCTION

1. "Battle of Megiddo (15th Century BC)," *Wikipedia*, https://en.wikipedia.org/wiki/Battle_of_Megiddo_(15th_century_BC) (accessed July 1, 2013); Jimmy Dunn, "The Battle of Megiddo," Tour Egypt, http://www.touregypt.net/featurestories/megiddo.htm (accessed July1, 2013).

2. N. S. Gill, "Pharaoh Thutmose III and the Battle of Megiddo," About.com, http://ancienthistory.about.com/od/egyptmilitary/qt/070607Megiddo.htm (accessed July 2, 2013).

CHAPTER 2. EARLY WARS AND THE BEGINNING OF PHYSICS

1. For an excellent account of the Battle of Kadesh, see Robert Collins Surh, "Battle of Kadesh, *Military History*, August 1995. The article can also be found online at Historynet.com, http://www.historynet.com/battle-of-Kadesh.htm (accessed July 23, 2013).

2. Ernest Volkman, *Science Goes To War* (New York: John Wiley and Sons, 2002), p. 17.

3. Ibid., p. 20.

4. Robert O'Connell, *Of Arms and Men* (New York: Oxford University Press, 1989), p. 39.

5. "Aristotle," Ancient Greece, http://www.ancientgreece.com/s/People/Aristotle (accessed December 15, 2012).

6. Linda Alchin, "Ballista," Middle Ages, http://www.middle-ages.org.uk/ballista.htm (accessed December 18, 2012).

7. Linda Alchin, "Trebuchet," http://www.middle-ages.org.uk/trebuchet.htm (accessed December 20, 2012).

8. W. W. Tarn, *Philip of Macedon, Alexander the Great* (Boston: Beacon Press, 1972).

9. Volkman, *Science Goes To War*, p.30

10. E. J. Dijksterhuis, *Archimedes* (Princeton, NJ: Princeton University Press, 1983).

CHAPTER 3. BASIC PHYSICS OF EARLY WEAPONS

1. Isaac Asimov, *The History of Physics* (New York: Walker, 1966), p. 13.
2. Ibid., p. 26.
3. Ibid., p. 65.
4. Ibid., p. 84.
5. Barry Parker, *Science 101: Physics* (Irvington, NY: Collins-Smithsonian, 2007), p. 24.
6. Ibid., p. 26.
7. "The Physics of Archery," Mr. Fizzix, 2001, http://www.mrfizzix.com/archery (accessed January 3, 2013).
8. Franco Normani, "The Physics of Archery," Real World Physics Problems, http://www.real-world-physics-problems.com/physics-of-archery.html (accessed January 5, 2013).

CHAPTER 4. THE RISE AND FALL OF THE ROMAN EMPIRE AND THE EARLY ENGLISH-FRENCH WARS

1. Ernest Volkman, *Science Goes to War* (New York: John Wiley and Sons, 2002), p. 35.
2. Robert O'Connell, *Of Arms and Men* (New York: Oxford University Press, 1989), p. 69.
3. "The Battle of Adrianople (Hadrianopolis)," Illustrated History of the Roman Empire, http://www.roman-empire.net/army/adrianople.html (accessed January 10, 2013).
4. David Ross, "The Battle of Hastings," Britain Express, http://www.britain express.com/History/battles/hastings.htm (accessed January 13, 2013).
5. Kennedy Hickman, "Hundred Years' War: Battle of Crécy," About.com, http://www.militaryhistory.about.com/od/battleswars12011400/p/crecy.htm (accessed January 16, 2013).
6. Volkman, *Science Goes to War*, p. 38.
7. "The Battle of Agincourt," BritishBattles.com, http://www.britishbattles.com/100-years-war/agincourt.htm (accessed January 19, 2013).
8. Robert Hardy, *Longbow: A Social and Military History* (New York: Lyons and Burford, 1993).
9. "The Physics of Archery," Mr. Fizzix, 2001, http://mrfizzix.com/archery (accessed January 21, 2013).
10. Franco Normani, "The Physics of Archery," Real World Physics Problems, http://www.real-world-physics-problems.com/physics-of-archery.html (accessed January 24, 2013).

CHAPTER 5. GUNPOWDER AND CANNONS:
THE DISCOVERIES THAT CHANGED
THE ART OF WAR AND THE WORLD

1. Jack Kelly, *Gunpowder* (New York: Basic Books, 2004), p. 12.
2. Ibid., p. 17.
3. J. R. Partington, *A History of Greek Fire and Gunpowder* (Baltimore: Johns Hopkins University Press, 1999), p. 22.
4. Kelly, *Gunpowder*, p. 23; Partington, *A History of Greek Fire*, p. 69.
5. Robert O'Connell, *Of Arms and Men* (New York: Oxford University Press, 1989), p. 108; Kelly, *Gunpowder*, p. 41.
6. Partington, *A History of Greek Fire*, p. 91.
7. "Huolongjing," *Wikipedia*, http://en.wikipedia.org/wiki/Huolongjing (accessed August 6, 2013).
8. Ernest Volkman, *Science Goes to War* (New York: John Wiley and Sons, 2002), p. 53; Kelly, *Gunpowder*, p .49.
9. Volkman, *Science Goes to War*, p. 63; Kelly, *Gunpowder*, p. 55.
10. Chris Trueman, "Charles VIII," History Learning Site, http://www.history learningsite.co.uk/c8.htm (accessed January 27, 2013).

CHAPTER 6. THREE MEN AHEAD OF THEIR TIME:
DA VINCI, TARTAGLIA, AND GALILEO

1. "Leonardo da Vinci," *Wikipedia*, http://en.wikipedia.org/wiki/leonardo_da _vinci (accessed January 29, 2013).
2. Christopher Lampton, "Top 10 Leonardo da Vinci Inventions," HowStuffWorks.com, January 25, 2011, http://www.howstuffworks.com/innovations/famous-inventors/ 10-Leonardo-da-Vinci-Inventions.htm (accessed February 1, 2013).
3. "Science and Inventions of Leonardo da Vinci," *Wikipedia*, http://en.wikipedia .org/Science_and_inventions_of_Leonardo_da_Vinci (accessed February 2, 2013)
4. Ernest Volkman, *Science Goes to War* (New York: John Wiley and Sons, 2002), p. 77.
5. "Tartaglia Biography," MacTutor History of Mathematics, http://www-history .mcs.st-and.ac.uk/Biographies/Tartaglia.html (accessed February 3, 2013).
6. J. Bronowski, *The Ascent of Man* (Boston: Little, Brown and Company, 1973), p. 198.
7. Mary Bellis, "Galileo Galilei," About.com, http://www.inventors.about.com/ od/gstartinventors/a/Galileo_Galilei.htm (accessed February 6, 2013); "Galileo Galilei," *Wikipedia*, http://en.wikipedia.org/wiki/Galileo_Galilei (accessed February 6, 2013).
8. This experiment was definitely performed a few years later.

CHAPTER 7. FROM EARLY GUNS TO
TOTAL DESTRUCTION AND DISCOVERY

1. J. R. Partington, *A History of Greek Fire and Gunpowder* (Baltimore: Johns Hopkins University Press, 1999), p. 97.

2. Jack Kelly, *Gunpowder* (New York: Basic Books, 2004), p. 70; "Matchlock," *Wikipedia*, http://en.wikipedia.org/wiki/matchlock (accessed February 9, 2013).

3. Kelly, *Gunpowder*, p. 76.

4. "Wheellock," *Wikipedia*, http://en.wikipedia.org/wiki/wheellock (accessed February 10, 2013).

5. Ernest Volkman, *Science Goes to War* (New York: John Wiley and Sons, 2002), p. 91; Matt Rosenberg, "Prince Henry the Navigator," About.com, http://www.geography.about.com/od/historyofgeography/a/princehenry.htm (accessed February 13, 2013).

6. Volkman, *Science Goes to War*, p. 99; F. Streicher, "Paolo dal Pozzo Toscanelli," *Catholic Encyclopedia* (New York: Robert Appleton, 1912), available online at New Advent, http://www.newadvent.org/cathen/14786a.htm (accessed February 14, 2013).

7. A. F. Pollard, "King Henry VIII," excerpted from *Encyclopedia Britannica*, 11th ed. (Cambridge: Cambridge University Press, 1910), 8: 289, available online at Luminarium.org, http://www.luminarium.org/renlit/tudorbio.htm (accessed February 15, 2013).

8. Mary Bellis, "William Gilbert," about.com, http://www.inventors.about.com/library/inventors/bl_william_gilbert.htm (accessed February 17, 2013).

9. Volkman, *Science Goes to War*, p. 104.

10. Dava Sobel, *Longitude: The True Story of a Lone Genius Who Solved the Greatest Problem of His Time* (New York: Walker and company, 2007).

11. Kelly, *Gunpowder*, p. 132.

12. R. L. O'Connell, *Of Arms and Men* (New York: Oxford University Press, 1989), p. 141.

13. "Gustavus Adolphus of Sweden," Answers.com, http://www.answers.com/topic/gustav-ii-adolph-of-sweden (accessed February 20, 2013).

14. Barry Parker, *Science 101: Physics* (Irvington, NY: Collins-Smithsonian, 2007), p. 8; J. Bronowski, *The Ascent of Man* (Boston: Little, Brown and Company, 1973), p. 221.

CHAPTER 8. THE IMPACT OF THE
INDUSTRIAL REVOLUTION

1. Ernest Volkman, *Science Goes to War* (New York: John Wiley and Sons, 2002), p. 116.

2. "Louis XIV Biography," Bio, http://www.biography.com/people/louis-xiv-9386885 (accessed February 22, 2013).

3. J. Bronowski, *The Ascent of Man* (Boston: Little, Brown and company, 1973), p. 259.

4. Mary Bellis, "James Watt—Inventor of the Modern Steam Engine," About.com, http://inventors.about.com/od/wstartinventors/a/james_watt.htm (accessed February 24, 2013); Carl Lira, "Biography of James Watt," Michigan State University College of Engineering, http://www.egr.msu.edu/~lira/supp/steam/wattbio.html (accessed February 24, 2013).

5. Bronowski, *Ascent of Man*, p. 274.

6. Volkman, *Science Goes to War*, p. 126; J. J. O'Connor and E. F. Robertson, "Benjamin Robins," MacTutor History of Mathematics Archive, http://www-history.mcs.st-andrews.ac.uk/Biographies/Robins.html (accessed February 27, 2013).

7. "Flintlock," *Wikipedia*, http://en.wikipedia.org/wiki/flintlock.

8. C. D. Andriesse, *Huygens: The Man behind the Principle* (Cambridge: Cambridge University Press, 2011).

9. "Christiaan Huygens," *Wikipedia*, http://en.wikipedia.org/wiki/Christiaan_Huygens (accessed March 1, 2013).

CHAPTER 9. NAPOLEON'S WEAPONS AND NEW BREAKTHROUGHS IN PHYSICS

1. Ernest Volkman, *Science Goes to War* (New York: John Wiley, 2002), p. 136.

2. J. Bronowski, *The Ascent of Man* (Boston: Little, Brown and company, 1973), p. 148.

3. "Antoine Lavoisier," *Wikipedia*, http://en.wikipedia.org/wiki/Antoine_Lavoisier (accessed August 15, 2013).

4. John H. Lienhard, "No. 728: Death of Lavoisier," Engines of Our Ingenuity, http://www.uh.edu/engines/epi728.htm (accessed August 15, 2013).

5. Robert Wilde, "Napoleon Bonaparte," About.com, http://europeanhistory.about.com/od/bonapartenapoleon/a/bionapoleon.htm (accessed March 3, 2013).

6. "Napoleonic Weaponry and Warfare," *Wikipedia*, http://en.wikipedia.org/wiki/napoleonic_weaponry_and_warfare (accessed March 3, 2013).

7. "French Invasion of Russia," *Wikipedia*, http://en.wikipedia.org/wiki/French_invasion_of_Russia (accessed March 4, 2013)

8. Woburn Historical Commission, "Count Rumford," Middlesex Canal website, http://www.middlesexcanal.org/docs/rumford.htm (accessed March 5, 2013).

9. Barry Parker, *Science 101: Physics* (Irvington, NY: Collins-Smithsonian, 2007), p. 110.

10. Ibid., p. 112.

11. Ibid., p. 116.

12. Ibid., p. 118.

CHAPTER 10. THE AMERICAN CIVIL WAR

1. Jack Kelly, *Gunpowder* (New York: Basic Books, 2004), p. 180.
2. R. L. O'Connell, *Of Arms and Men* (New York: Oxford University Press, 1989), p. 191.
3. Kelly, *Gunpowder*, p. 182.
4. Ibid., p. 188; O'Connell, *Of Arms and Men*, p. 191.
5. Kelly, *Gunpowder*, p. 213; O'Connell, *Of Arms and Men*, p. 196.
6. "American Civil War," *Wikipedia*, http://en.wikipedia.org/wiki/american_civil_war (accessed March 9, 2013).
7. "Battle of Gettysburg," The History Place, http://www.historyplace.com/civilwar/battle.htm (accessed March 9, 2013).
8. Howard Taylor, "The Telegraph in the War Room," Learning-Online, http://www.alincolnlearning.us/Civilwartelegraphing.html (accessed March 13, 2013).
9. Mary Bellis, "Introduction to Joseph Henry," About.com, http://inventors.about.com/od/hstartinventors/a/Joseph_Henry.htm (accessed March 14, 2013).
10. Barry Parker, *Science 101: Physics* (Irvington, NY: Collins-Smithsonian, 2007), p. 118.
11. "Gatling Gun," *Wikipedia*, http://en.wikipedia.org/wiki/gatling_gun (accessed March 16, 2013).
12. Kelly, *Gunpowder*, p. 191.
13. Craig L. Symonds, "Damn the Torpedoes! The Battle of Mobile Bay," Civil War Trust, http://www.civilwar.org/battlefields/mobilebay/mobile-bay-history-articles/damn-the-torpedoes-the.html (accessed March 18, 2013).
14. "Civil War Submarines," AmericanCivilWar.com, http://americancivilwar.com/tcwn/civil_war/naval_submarine.html (accessed March 19, 2013).
15. "Balloons in the American Civil War," CivilWar.com, http://www.civilwar.com/weapons/observation_balloons.html (accessed March 21, 2013).

CHAPTER 11. WHERE DOES THE BULLET GO? BALLISTICS OF RIFLE BULLETS AND CANNON SHELLS

1. "Internal Ballistics," *Wikipedia*, http://en.wikipedia.org/wiki/internal_ballistics (accessed March 24, 2013).
2. Nelson DeLeon, "Elementary Gas Laws: Charles Law," Chemistry 101 Class Notes, Spring 2001, http://www.iun.edu/~cpanhd/C101webnotes/gases/charleslaw.html (accessed March 25, 2013).
3. "Recoil," *Wikipedia*, http://en.wikipedia.org/wiki/recoil (accessed March 27, 2013).
4. "Introduction to Ballistics," Federation of American Scientists, http://www.fas.org/man/dod-101/navy/docs/swos/gunno/INFO6.html (accessed March 29, 2013).

5. "External Ballistics," *Wikipedia*, http://en.wikipedia.org/wiki/external_ballistics (accessed April 1, 2013).

6. "Terminal Ballistics," *Wikipedia*, http://en.wikipedia.org/wiki/terminal_ballistics (accessed April 1, 2013).

CHAPTER 12. HEY, LOOK . . . IT FLIES! AERODYNAMICS AND THE FIRST AIRPLANES

1. Isaac Asimov, *The History of Physics* (New York: Walker and Company, 1966), p. 133.

2. "Wright Brothers History: First Airplane Flight," Welcome to the Wright House, http://www.wright-house.com/wright-brothers/wrights/1903.html (accessed April 5, 2013).

3. Mary Bellis, "A Visual Timeline: The Lives of the Wright Brothers and Their Invention of the Airplane," About.com, http://inventors.about.com/od/wstartinventors/a/TheWrightBrothers.htm (accessed April 5, 2013).

4. Quentin Reynolds, *The Wright Brothers: Pioneers of American Aviation* (New York: Random House, 1981).

5. Fred Howard, *Wilbur and Orville: A Biography of the Wright Brothers* (New York: Ballantine Books, 1988), p. 72.

6. "What Makes an Airplane Fly—Level 1," Allstar Network, http://www.allstar.fiu.edu/aero/fltmidfly.htm (accessed April 8, 2013).

7. Mary Bellis, "The Dynamics of Airplane Flight," About.com, http://inventors.about.com/library/inventors/blairplanedynamics.htm (accessed April 8, 2013).

8. "What Is Drag?" National Aeronautics and Space Administration, http://www.grc.nasa.gov/WWW/k-12/airplane/drag1.html (accessed April 10, 2013).

9. "The Birth of the Fighter Plane, 1915," EyeWitness to History, 2008, http://www.eyewitnesstohistory.com/fokker.htm (accessed April 14, 2013).

10. "Aviation in World War I," *Wikipedia*, http://en.wikipedia.org/wiki/Aviation_in_World_War_I (accessed April 14, 2013); R. L. O'Connell, *Of arms and Men* (New York: Oxford University Press, 1989), p. 262.

CHAPTER 13. THE MACHINE GUN WAR—WORLD WAR I

1. Ernest Volkman, *Science Goes to War* (New York: John Wiley, 2002), p. 151; R. L. O'Connell, *Of Arms and Men* (New York: Oxford University Press, 1989), p. 233.

2. Michael Duffy, "Weapons of War—Machine Guns," firstworldwar.com, http://www.firstworldwar.com/weaponry/machineguns.htm (accessed April 20, 2013).

3. "World War I—Weapons," History on the Net, http://www.historyonthenet.com/WW1/weapons.htm (accessed April 22, 2013).

4. Michael Duffy, "How It Began—Introduction," firstworldwar.com, http://www.firstworldwar.com/origins/ (accessed April 25, 2013); Jennifer Rosenberg, "World War I, About.com, http://history1900s.about.com/od/worldwari/p/World-War-I.htm (accessed April 28, 2013).

5. O'Connell, *Of Arms and Men*, p. 262.

6. Stephen Sherman, "Legendary Aviators and Aircraft of World War One," 2001, Acepilots.com, http://acepilots.com/wwi/ (accessed April 30, 2013).

7. Michael Duffy, "The War in the Air—Air Aces of World War One," firstworldwar.com, http://www.firstworldwar.com/features/aces.htm (accessed April 30, 2013).

8. "Jan. 31, 1917: Germans Unleash U-Boats," This Day in History, History, http://www.history.com/this-day-in-history/germans-unleash-u-boats (accessed May 3, 2013); Alex L., "U-Boats in World War I," HistoryJournal.org, http://historyjournal.org/2012/08/28/u-boats-in-world-war-i/ (accessed May 5, 2013).

9. "The Sinking of the RMS *Lusitania*," *Wikipedia*, http://en.wikipedia.org/wiki/Sinking_of_the_RMS_Lusitania (accessed May 4, 2013).

10. "Poison Gas and World War One," History Learning Site, http://www.historylearningsite.co.uk/poison_gas_and_world_war_one.htm (accessed May 5, 2013).

11. Michael Duffy, "Weapons of War—Poison Gas," firstworldwar.com, http://www.firstworldwar.com/weaponry/gas.htm (accessed May 7, 2013).

12. "Chemical Weapons in World War I, *Wikipedia*, http://en.wikipedia.org/wiki/Chemical_weapons_in_World_War_I (accessed September 5, 2013).

13. "Tanks and World War One," History Learning Site, http://www.historylearningsite.co.uk/tanks_and_world_war_I (accessed May 11, 2013).

14. Michael Duffy, "Weapons of War—Tanks," firstworldwar.com, http://www.firstworldwar.com/weaponry/tanks.htm (accessed May 11, 2013).

15. "Apr. 6, 1917: America Enters World War I," This Day in History, History, http://www.history.com/this-day-in-history/america-enters-world-war-i (accessed May 12, 2013).

CHAPTER 14. THE INVISIBLE RAYS: THE DEVELOPMENT AND USE OF RADIO AND RADAR IN WAR

1. Barry Parker, *Science 101: Physics* (Irvington, NY: Collins-Smithsonian, 2007), p. 129.

2. Ibid., p. 122.

3. Ibid., p. 121; "Guglielmo Marconi," *Wikipedia*, http://en.wikipedia.org/wiki/Guglielmo_Marconi (accessed May 15, 2013).

4. Parker, *Science 101*, pp. 123, 132.

5. "Learn about Australian Weather Watch Radar," Australian Government Bureau of Meteorology," http://www.bom.gov.au/australia/radar/about (accessed May 17, 2013).

6. Robert Buderi, *The Invention That changed the World* (New York: Simon and Schuster, 1996), p. 103.

7. Louis Brown, *A Radar History of World War II* (Philadelphia: Institute of Physics Publishing, 1999), p. 84.

8. James Phinney Baxter III, quoted in Buderi, *Invention That Changed the World*.

CHAPTER 15. SONAR AND THE SUBMARINE

1. Isaac Asimov, *The History of Physics* (New York: Walker and Company, 1966), p. 124.

2. Nathan Earls, "The Physics of Submarines," University of Alaska Fairbanks, http://ffden-2.phys.uaf.edu/212_fall2003.web.dir/nathan_earls/intro_slide.html (accessed May 20, 2013).

3. Marshall Brain and Craig Freudenrich, "How Submarines Work," How Stuff Works," http://science.howstuffworks.com/transport/engines-equipment/submarine (accessed May 22, 2013).

4. "Sonar," *Wikipedia*, http://en.wikipedia.org/wiki/sonar (accessed May 25, 2013).

5. Mary Bellis, "The History of Sonar," About.com, http://inventors.about.com/od/sstartinventions/a/sonar_history.htm (accessed May 25, 2013).

6. "Torpedo," *Wikipedia*, http://en.wikipedia.org/wiki/torpedo (accessed May 28, 2013).

7. "The German U-Boats," uboat.net, http://www.uboat.net/boats.htm (accessed May 30, 2013).

8. "Battle of the Philippine Sea," *Wikipedia*, http://en.wikipedia.org/wiki/Battle_of_the_Philippine_Sea (accessed May 30, 2013).

CHAPTER 16. THE GREAT WAR: WORLD WAR II

1. Jennifer Rosenberg, "World War II Starts," About.com, http://history1900s.about.com/od/worldwarii/a/wwiistarts.htm (accessed June 1, 2013); "World War Two—Causes," History on the Net.com, http://www.historyonthenet.com/WW2/causes.htm (accessed June 1, 2013).

2. Editors of Legacy Publishers, "Start of World War II: September 1939–March 1940," How Stuff Works, http://history.howstuffworks/world-war-ii/start-world-war-2.htm (accessed June 1, 2013).

3. TheophileEscargot, "1940; The Battle of France," Kuro5hin, http://www.kuro5hin.org/story/2002/5/14/55627/2665 (accessed June 4, 2013).

4. Louis Brown, *A Radar History of World War II* (Philadelphia: Institute of Physics, 1979).

5. Robert Buderi, *The Invention That Changed the World* (New York: Simon and Schuster, 1996), p. 79.

6. Brown, *Radar History*, p. 107.

7. Buderi, *Invention*, p. 89.

8. "The Battle of Britain," BBC, http://www.bbc.co.uk/history/battle_of_britain (accessed June 5, 2013).

9. "Junkers Ju 87," *Wikipedia*, http://en.wikipedia.org/wiki/Junkers_Ju_87 (accessed June 6, 2013).

10. "Reasons for America's Entry into WWII," Hubpages, http://jdf78.hubpages.com/hub/Reasons-for-American-Entry-Into-WWII (accessed June 7, 2013).

11. "Air Warfare of World War II," *Wikipedia*, http://en.wikipedia.org/wiki/Air_warfare_of_World_War_II (accessed June 8, 2013).

12. "V-2 Rocket," *Wikipedia*, http://en.wikipedia.org/wiki/V-2_rocket (accessed June 9, 2013); Kennedy Hickman, "World War II: V-2 Rocket," About.com, http://militaryhistory.about.com/od/artillerysiegeweapons/p/v2rocket.htm (accessed June 10, 2013).

13. "Norden Bombsight, *Wikipedia*, http://en.wikipedia.org/wiki/Norden_bombsight (accessed June 11, 2013).

14. "World War 2 Code Breaking: 1939–1945," History, http://www.history.co.uk/explore-history/ww2/code-breaking.html (accessed June 13, 2013).

15. "More Information About: Alan Turing," BBC, http://www.bbc.co.uk/history/people/alan_turing (accessed June 14, 2013).

CHAPTER 17. THE ATOMIC BOMB

1. Isaac Asimov, *The History of Physics* (New York: Walker and Company, 1966), p. 598.

2. Amir Aczel, *Uranium Wars* (New York: MacMillan, 2009), p. 179.

3. Ibid., p. 74.

4. Ibid., p. 88.

5. Richard Rhodes, *The Making of the Atomic Bomb* (New York: Simon and Schuster, 1986), p. 204.

6. Ibid., p. 79.

7. Aczel, *Uranium Wars*, p. 61.

8. Ibid., p. 104.

9. Rhodes, *Making of the Atomic Bomb*, p. 256.

10. Barry Parker, *Quantum Legacy* (Amherst, NY: Prometheus Books, 2002), p. 217.

11. Ibid., p. 213.

12. Aczel, *Uranium Wars*, p. 132.

13. Jim Baggott, *The First War of Physics* (New York: Pegasus, 2010), p. 100.

14. Ibid., p. 89; Aczel, *Uranium Wars*, p. 146.
15. Baggott, *First War of Physics*, p. 232.
16. Rhodes, *Making of the Atomic Bomb*, p. 447.
17. Aczel, *Uranium Wars*, p. 157.
18. Baggott, *First War of Physics*, p. 279.
19. Ibid., p. 299.
20. Aczel, *Uranium Wars*, p. 178.

CHAPTER 18. THE HYDROGEN BOMB, INTERCONTINENTAL MISSILES, LASERS, AND THE FUTURE

1. "Thermonuclear Weapon," *Wikipedia*, http://en.wikipedia.org/wiki/Thermonuclear_weapon (accessed June 20, 2013).
2. Richard Rhodes, *Dark Sun* (New York: Simon and Schuster, 1995), p. 466; "Cold War: A Brief History of the Atomic Bomb," atomicarchive.com, http://www.atomicarchive.com/history/coldwar/page04.htm (accessed June 22, 2013).
3. Rhodes, *Dark Sun*, p. 468; "Thermonuclear Weapon."
4. Rhodes, *Dark Sun*, p. 482.
5. Ibid., p. 506.
6. "Intercontinental Ballistic Missile," *Wikipedia*, http://en.wikipedia.org/wiki/intercontinental_ballistic_missile (accessed June 26, 2013).
7. Barry Parker, *Quantum Legacy* (Amherst, NY: Prometheus Books, 2002), p. 159.
8. "Gordon Gould," *Wikipedia*, http://en.wikipedia.org/wiki/Gordon_Gould (accessed September 12, 2013).
9. Parker, *Quantum Legacy*, p. 179.
10. "Extrinsic Semiconductor, *Wikipedia*, http://en.wikipedia.org/wiki/Extrinsic_semiconductor (accessed September 12, 2013).
11. "What Is the Keyhole Satellite and What Can It Really Spy On?" How Stuff Works, http://science.howstuffworks.com/question529.htm (accessed September 12, 2013).
12. "Unmanned Aerial Vehicle, *Wikipedia*, http://en.wikipedia.org/wiki/Unmanned_aerial_vehicle (accessed June 27, 2013).
13. Joe Haldeman and Martin Greenberg, *Future Weapons of War* (Riverdale, NY: Baen, 2008).
14. "How E-Bombs Work," How Stuff Works, http://sciencehowstuffworks.com/e-bomb3.htm (accessed June 28, 2013).
15. Joel Baglole, "XM25- Future Grenade Launcher," About.com, http://usmilitary.about.com/od/weapons/a/xm25grenadelaunch.htm (accessed June 29, 2013).
16. "Smart Bullet," *Wikipedia*, http://en.wikipedia.org/wiki/Smart_bullet (accessed September 12, 2013).
17. "MAARS," Qinetiq North America, https://www.qinetiq-na.com/products/unmanned-systems/maars/, (accessed September 12, 2013).

SELECTED BIBLIOGRAPHY

Aczel, Amir. *Uranium Wars*. New York: MacMillan, 2009.

Asimov, Isaac. *The History of Physics*. New York: Walker, 1966.

Baggott, Jim. *The First War of Physics*. New York: Pegasus, 2010.

Bronowski, J. *The Ascent of Man*. Boston: Little, Brown, 1973.

Brown, Louis. *A Radar History of World War II*. Philadelphia: Institute of Physics Publishing, 1999.

Buderi, Robert. *The Invention That Changed the World*. New York: Simon and Schuster, 1996.

Collier, Basil. *The Battle of Britain*. New York: MacMillan, 1962.

Griffith, Paddy. *Battle Tactics of the Civil War*. New Haven, CT: Yale University Press, 1987.

Guillen, Michael. *Five Equations That Changed the World*. New York: Hyperion, 1996.

Guilmartin, John. *Gunpowder and Galleys*. Cambridge: Cambridge University Press, 1964.

Hardy, Robert. *Longbow: A Social and Military History*. New York: Lyons and Burford, 1993.

Hodges, Andrew. *Alan Turing: The Enigma*. Princeton, NJ: Princeton University Press, 2012.

Hughes, B. P. *Firepower: Weapon Effectiveness on the Battlefield, 1630–1850*. New York: Da Capo, 1997.

Jones, R. V. *Most Secret War*. New York: Penguin, 2009.

Keegen, John. *A History of Warfare*. New York: Vintage, 1994.

Kelly, Jack. *Gunpowder*. New York: Basic Books, 2004.

Kennedy, Gregory. *Germany's V-2 Rocket*. Atglen, PA: Schiffer, 2006.

Maraden, E. W. *Greek and Roman Artillery*. New York: Oxford, 1969.

O'Connell, Robert. *Of Arms and Men*. New York: Oxford, 1989.

Padfield, Peter. *Guns at Sea*. New York: St. Martin's, 1974.

Pais, Abraham. *J. Robert Oppenheimer: A Life*. New York: Oxford University Press, 2006.

Parker, Barry. *Science 101: Physics*. Irvington, NY: Collins, 2007.

Partington, J. R. *A History of Greek Fire and Gunpowder*. Baltimore: Johns Hopkins University Press, 1999.

Rhodes, Richard. *Dark Sun*. New York: Simon and Schuster, 1995.

———. *The making of the Atomic Bomb*. New York: Simon and Schuster, 1986.

Sarton, George, *Ancient Science through the Golden Age of Greece*. Mineola, NY: Dover Publications, 2011.

Sebag-Montefiore, Hugh. *Enigma: The Battle of the Code*. New York: Wiley, 2004.

Snodgrass, A. M. *Arms and Armor of the Greeks*. Baltimore: Johns Hopkins University Press, 1998.

Volkman, Ernest. *Science Goes to War*. New York: Wiley, 2002.

Weller, Jac. *Weapons and Tactics*. Boulder, CO: Paladin Press, 2007.

INDEX